MASTERING LOGICAL FALLACIES

For general information on our other products and services or to obtain technical support, please contact our Customer Care Department within the United States at (866) 744-2665, or outside the United States at (510) 253-0500.

Zephyros Press publishes its books in a variety of electronic and print formats. Some content that appears in print may not be available in electronic books, and vice versa.

Illustrations © 2016 by James Olstein.

ISBN: Print 978-1-62315-710-4 | eBook 978-1-62315-711-1

MASTERING LOGICAL FALLACIES

The Definitive Guide to Flawless Rhetoric and Bulletproof Logic

MICHAEL WITHEY

ILLUSTRATIONS BY JAMES OLSTEIN

ZEPHYROS PRESS

Contents

THE FALLACIES

Foreword

I'VE ALWAYS THOUGHT OF ARGUMENTS as a game of chess—there are rules, and beyond those, there are countless moves you might make, some more strategic than others. The worst players, then, those that fall for cheap tricks, commit avoidable blunders and thus are easily beaten, are those without a good understanding of these rules and tactics. If I have learned anything in ten years of formal debating, it is that arguments are no different: without a good understanding of the rules and tactics, you are likely to do poorly and be beaten.

Some things are patently true and others obviously false. Most people can spot those immediately. As it happens, everything else in the world is not this clear, especially when it comes to the complex matters that policymakers, business leaders, and academics agonize over: How much domestic surveillance on our government's behalf is acceptable? How should economic development proceed, when it is at the expense of the environment? What kind of duty, if any, do children have towards their parents? What is it I personally and we, collectively, should want out of life?

Mastering chess and mastering arguments do differ, then, in one very important respect: not everyone must learn to play chess, but everyone, without exception, will encounter arguments. Learning the rules and tactics of debating (the subject of this book) is nothing less than an essential life skill, like learning proper grammar, or being able to cook. True, you might still be able to survive without becoming a master of any of these, but your life will likely be measurably compromised.

How is it, exactly, that mastering arguments makes your life better? To start with, it helps help you find answers to questions of political, economics, academic, and personal import. Fundamentally, these questions ask you to weigh, often speculatively, many different, and frequently opposing, issues and value propositions. In other words, they clearly leave room for argument, so it is no surprise that the best answers come from those with the most convincing grasp of the arguments.

Perhaps, above all, you want to avoid being fooled by someone else's logical sleight of hand. The fact that this book can provide a long list of real-life examples of abuses of reasons indicates just how likely it is that you may be won over by one, or that you already have. Everyone, from our relatives to our coworkers to our elected officials, has a vested interest in convincing us about theories that may be false, or against our own interest, because it is in *their* interest to do so. How will you be able to hold these people accountable, if you cannot realize you are being deceived in the first place?

Awareness, then, is the key, and that is precisely the skill this book teaches. Like other life skills, recognizing when abuses of logic are being perpetrated eventually becomes second nature–you just *know* it when something seems fishy about an argument. And, like the other skills, everyone must start with an instructional manual before they can eventually become proficient.

Ultimately, this proficiency will refine your dealings with others in two ways. First, when a logical tresspass occurs, you will not be fooled! Instead, you will be able to point it out. A devastating comeback is, as we know, often the end of an argument. Sometimes, it can add a gloss of authority to your comeback if you use the precise name of a fallacy, but the majority of everyday conversations probably command a more subtle approach. This book makes particular note of how you might appropriately craft your response.

Secondly, and most importantly, you'll find you construct better arguments of your own (and avoid being on the receiving end of a stinging comeback). Recognizing the vulnerabilities of others' arguments means recognizing your potential vulnerabilities as well. This is incredibly useful because in your everyday life you are constantly expected to convince someone of something: an interviewer, a teacher, a sibling, a friend, etc.

Finally, you will be able to better answer all those questions of public and private importance for yourself, rather than be seduced by someone else's word for it. The best thinkers are the best arguers, defenders of their point of view, and there is no reason you can't be one of them. In this respect, arguments are indeed like chess—watching the best is beautiful, if only because their brilliance is obvious.

Have I made a sound argument for studying the contents of this book? There is one way to find out: master the rules and tactics of arguments. As a friend would, I cannot recommend more strongly that you do.

–Henry Zhang

Introduction

ARGUMENTS ARE EVERYWHERE: turn on your TV, flick through the channels, and you will, no doubt, come across some demagogue giving a spiel about why you should vote for him, rather than some other schlub. Break to commercials, and an advertisement will promise you that, if you use this or that deodorant, women will fall at your feet. Turn off your TV, have dinner with your family: no doubt your uncle will start telling you about why you should fix your engine gasket, or who to vote for in the next election.

Arguments are, indeed, everywhere. The trouble is that a lot of arguments are actually really bad. In a good argument, the conclusion follows from the premises; in a bad argument, it doesn't. Something has gone wrong: the premises are irrelevant to the conclusion, or the conclusion is much stronger than the premises allow, or, even worse, the inference drawn requires violating basic rules of logic. But here's the thing: experience shows that bad arguments are *effective*. Even though their conclusions don't follow from their premises, people still believe them. Hence, people will hold false and pernicious beliefs, because they have been fooled by faulty arguments. This book will help you be on your guard against such arguments. It will help you recognize them, and give you some rebuttals to use when you encounter them.

When an argument goes wrong, it's because its proponent has commited a *fallacy*. Now, fallacies come in two main kinds: *formal* and *informal*. A formal fallacy is one where the argument is false simply by virtue of its form. We can think of this sort of fallacy as one with a faulty *structure*: the argument tries to infer the conclusion from its premises; it cannot do so, however, because of the argument's *structure*. These arguments should always be rejected, because they violate the fundamental rules of logic. So, if someone says, "All pigeons are birds; Tweety is a bird; therefore Tweety is a pigeon," the argument is *formally* fallacious: Tweety could be a canary, an eagle, or a cormorant.

An informal fallacy is a trickier beast: it is one where the conclusion doesn't follow from the premises not because of the logical structure of the argument, but because of the *content* of the premises and the conclusion. We cannot, therefore, reject its conclusion simply because of the *form* of the argument; rather, we must look and see what the premises and conclusion actually are. It is harder to diagnose and rebut an informal fallacy, because you have to look not only at the content of the premises and conclusion, but also the use to which the argument is put.

So, if a proponent argues, "There are men on the moon, because there are men on the moon," his argument is logically *valid*: the conclusion does indeed follow from the premises, simply because the premises are *identical* to the conclusion. The argument is clearly circular and, therefore, worthless. But that doesn't mean you *should* reject it. If you don't believe there are men on the moon, this argument won't convince you; if you, however, accept the premises, then you have to accept the argument, no matter how trivial it may be. So there isn't, as with formal fallacies, a hard-and-fast rule that says you *must* reject arguments of this sort; rather, you have to check their validity on a case-by-case basis.

The vast majority of fallacies in this book are of the informal variety. Some of these fallacies are simply rhetorical techniques that appeal to feelings rather than reason. So, when a demagogue whips up a crowd into anger, or plays on their fear of some minority, his argument simply tries to bypass our reason and elicit an emotional reaction. Such fallacies are usually quite easy to spot. However, other informal fallacies go wrong for more subtle reasons: they play on words; they generalize from examples that are too limited; they falsely shift the burden of proof. Recognizing, and rebutting, these arguments is a lot trickier.

Although it is hard–actually, impossible–to lay down clear-cut rules on these matters, reading this book will tell you what fallacies you should look out for, and provide you with useful tips on countering your opponent's arguments. The best way, of course, to completely master logical fallacies is by long and hard practice: you must remain vigilant and constantly think about the arguments you encounter, and then try and work out what may have gone wrong.

It's a practice that is definitely worth undertaking. There are simply too many bad arguments out there, and their effect is by no means negligible. They lead us to believe pernicious and false ideas, and perpetuate terrible decisions, both on an individual and a collective level. It may take time and effort to cure all that, but the book you're holding in your hands is undeniably the best possible place to start.

Key Terms

A list of the most common terms you'll encounter in this book

ARGUMENT A set of statements (premises) that are given to advance a conclusion.

AUDIENCE The people listening to the argument, those to whom the argument is directed.

COMMITMENT A proposition that a proponent takes to be true, either implicitly or explicitly.

CONCLUSION A statement that follows from the preceding statements (the argument).

CONDITIONAL A statement of the form 'if A, then B,' where A cannot be true and B false.

DEDUCTIVE ARGUMENT An argument whose conclusion follows logically from its premises.

FALLACY The use of invalid reasoning in an argument.

INDUCTIVE ARGUMENT An argument whose premises give good evidence for its conclusion (e.g., inferring a general law from particular examples of it).

INVALID An argument is invalid only when it is not valid.

LOGIC The principles and rules that govern valid inferences.

MODAL A type of statement that deals with necessity, possibility or impossibility.

OPPONENT A person attempting to rebut an argument.

PRESUPPOSITION A proposition which an argument, or a proponent, takes for granted.

PROPONENT The person putting forward an argument.

SAMPLE A subset of a population, studied to make extrapolations that apply to the population as a whole.

SOUND An argument is sound only when it is valid and its premises are true.

VALID An argument is valid *if and only if* its premises cannot be true when its conclusion is false.

> **P1,2, ETC.** This symbol means "Statement 1, 2, etc."
>
> ∴ This symbol means "Therefore"

The
FALLACIES

AD HOMINEM: ABUSIVE

Informal	Person A makes claim *P*; person B states that A has a bad character; therefore, *P* is false.

Attacking a speaker's argument by insulting the speaker.

EXAMPLE

> *"You say that the Earth goes 'round the Sun: but you are a drunk and a womanizer!"*

REAL-LIFE EXAMPLE

Cicero, the Roman orator, was of humble origins, but rose to become the pre-eminent politician, lawyer and public speaker of his day. Despite his brilliance, however, the ruling aristocratic elite sometimes looked down on Cicero, precisely because he came from a modest background. We find an especially appalling example of this disdain in Cicero's dispute with Metellus Nepos who, in arguing with Cicero in a court of law, repeatedly asked "Cicero, who's your father?" Metellus Nepos couldn't argue against Cicero's case, and so attacked Cicero for his humble origins. Attacks on the race, class, or gender of the speaker are especially heinous examples of arguments *ad hominem*.

THE MISTAKE

An *ad hominem* argument is simply *irrelevant*: it doesn't contribute anything to the original argument itself, it only says something about the person making it (although it says rather more about the abuser employing the ad hominem). But the character of the

person making an argument doesn't affect the truth of the argument, or the validity of the inference. So it is with the chap who claims the Earth goes around the Sun: whether he's a drunk or a teetotaler, a womanizer or celibate, neither affect the truth of his claim. (As Galileo put it in a different context: "Still, it moves.")

THE COMEBACK

In an ideal world, there would be no need for a comeback: if the opponent has to resort to personal attacks, it should be clear that he's got nothing to say against the argument itself. In an ideal world, you should be able to win the argument just be pointing this out.

Alas, we do not live in this ideal world: ad hominem attacks are far more effective than worthy. One could retort with an ad hominem in turn, but this would just be stooping to your opponent's level. (However, it is worth noting Cicero's reply to Metellus Nepos's question "Cicero, who's your father?": "In your case, your mother has made the answer to this question rather difficult.").

SIGNIFICANCE

The forms of ad hominem argument we've discussed are sometimes called *ad personam* (against the *person*), to distinguish them from ad hominem attacks directed at the *commitments* of the speaker. The examples we have discussed above are clearly fallacious, since they are irrelevant to the argument at hand.

AD HOMINEM: CIRCUMSTANTIAL

Argument ex Concessis; Appeal to Motive; Vested Interest.

Informal	Person A claims that *P*. The circumstances of A discredit his assertion that *P*. Hence, we should disbelieve *P*.

Undermining the credibility of an argument by appealing to some facts about its proponent, where these facts are inconsistent with the proponent's advocacy of the argument, or where they undermine the proponent's credibility in putting forward the argument.

EXAMPLE

> *"The CEO of Oil America claims that drilling in Alaska will have a negligible environmental effect. But we shouldn't believe him: he's just saying that to get permission to drill!"*

REAL-LIFE EXAMPLE

We may cite Mandy Rice-Davies giving evidence in the Profumo Trial (one of the greatest political scandals of post-war Britain) regarding her affair with statesman Lord Astor. Pressed by the defense counsel, who pointed out that Lord Astor denied ever having met her, Rice-Davies delivered the immortal riposte: "Well, he would, wouldn't he?"

A more serious example is provided by the debate regarding vaccination. Lots of parents in Britain and America refuse to vaccinate their children against deadly infectious diseases, fearing that vaccination damages children, causing autism and bowel disease. When asked why the evidence shows overwhelmingly that vaccines don't actually have these adverse effects, they typically appeal to the vested interest of the pharmaceutical companies putting forward the evidence: *of course* they would cover up the adverse effects of the vaccines; if they didn't, they would lose their market. This move is common to a lot of conspiracy theorists.

THE MISTAKE

Like any *ad hominem*, this kind of argument is simply irrelevant: it does not attack the argument itself, but only its proponent. However, the argument is either true or false regardless of who its proponent is: the claim that "drilling in Alaska will have a negligible environmental effect" is just as true (or false, depending on the facts) whether made by the CEO of Oil America or a Greenpeace activist.

THE COMEBACK

Like the abusive *ad hominem* (see p. 16), one can simply point out that this response is irrelevant: the response should be directed at the argument (*ad rem*), not its proponent's possible motives.

But this move won't always work. In cases where the proponent is giving testimony, his circumstances can be relevant. So, Lord Astor's testimony that he never met Mandy Rice-Davies is clearly untrustworthy, since he would not want to admit that he consorted with an escort. Similarly, we can't just accept the CEO's assertion that Oil America's activities won't harm the environment, because he has a vested interest in getting permission to drill.

As such, if your opponent appeals to vested interest, you need to show that your argument can stand impartial scrutiny. Hence, you need to present the evidence for your claim and let your opponent try to find holes (if any) in it, or else appeal to a neutral third party to obtain impartial testimony in support of your claims.

Impugning the motives of an argument's proponent is one kind of circumstantial ad hominem; it's not the only one, however. Another kind questions whether the proponent can be committed to his argument given his other commitments, be they implicit or explicit. So, if the proponent argues that nothing can be known, you can ask him: "Are you *sure* about that?" Socrates's method of question-and-answer seems to proceed entirely by this method; indeed, some writers[1] have argued that *all* philosophical argument reduces, in the final analysis, to such circumstantial ad hominems.

---

1 Johnstone, *Philosophy and Argument.*

AD HOMINEM: GUILT BY ASSOCIATION

Informal	Opponent A argues that *P*. But a third party B also argues that *P*. B is unsavory. Hence, we should disbelieve that *P*. (Implicit premise: if B is unsavory, we should reject everything they say).
	──────── OR ────────
	The proponent of argument *P* associates with B. But B is unsavory. Hence, we should disbelieve *P*.

Attacking an argument by casting aspersions on people or organizations associated with either its proponent or the argument itself.

EXAMPLE

> "*John believes in higher taxation for the wealthy. But Castro also believed in higher taxes for the wealthy. Hence, we should reject higher taxes for the wealthy!*"

REAL-LIFE EXAMPLE

In 2008, Barack Obama (then a U.S. presidential candidate) came under fire for associating with pastor Jeremiah Wright, whose sermons contained controversial comments about September 11th and the Iraq War. Hillary Clinton, Bill O'Reilly and Mark Steyn all criticized Obama for this. John McCain, to his credit, defended this relationship, saying:

"I think that when people support you, it doesn't mean that you support everything they say. Obviously, those words and those statements are statements that none of us would associate ourselves with, and I don't believe that Senator Obama would support any of those, as well."

THE MISTAKE

This argument takes two forms, depending on whether it's the speaker or the argument that's presumed to be guilty by association. In the former case, it's clear that this is an especially foolish form of abusive *ad hominem* (see p. 16). The abusive *ad hominem* attacks the character of an argument's proponent, which is obviously irrelevant to the truth or falsity of his claims; therefore, if that move is fallacious, so too is attacking the proponent's associates.

A similar response can be made when the charge is leveled at the argument. The abusive *ad hominem* says: if the proponent of *P* is unsavory, we should reject *P*. This kind of argument says: if *P* has *any* proponent who is unsavory, we should reject *P*. But if it's irrelevant to the truth of *P* that its proponent is unsavory, its being embraced by a third party—however unsavory that party may be—is at least equally irrelevant to its truth.

THE COMEBACK

Again, one simply has to point out that the opponent's charge is *irrelevant*: what's at issue here is not my character, still less that of my friends; all that matters is the argument at hand. Similarly, it doesn't matter if some other odious person shares my beliefs: that in itself doesn't necessarily stop them from being true. The response, then, is simply to tell your opponent to stick to the matter at hand.

When it's the argument that is deemed guilty by association, this is very much a *Reductio ad Hitlerum* (see p. 168): "You believe that *P*, but Hitler also believed that *P*: so we should reject the view that *P*." But Hitler believed all sorts of things: that the earth goes around the sun, for instance. It doesn't follow that, just because Hitler believed that the earth goes around the sun, we should therefore reject the truth of this statement.

Nevertheless, this type of argument *may* in certain cases be a reasonable form of a circumstantial *ad hominem* (see p. 19), assuming that the associates of the argument's proponent reasonably cast doubt on his neutrality. So, if the proponent claims that Oil America's drilling will have a negligible environmental effect, his claim to neutrality could be compromised if it was revealed that he is best friends with Oil America's CEO. Whether this riposte is valid or not will, as in the case of the *Circumstantial ad Hominem*, depends on the circumstances.

AD HOMINEM: TU QUOQUE

Informal	The proponent makes an argument *P* against a certain behavior or action *Q*; but the proponent himself engages in *Q*. Hence, we should disbelieve *P*.

Undermining an argument against a certain behavior or action on the grounds that the proponent himself engages in the very same behavior or action.

EXAMPLE

"My dad always warns me against taking up smoking. But he gets through a pack a day! So why should I listen to his warning?"

REAL-LIFE EXAMPLE

The UK's Prime Minister, David Cameron, recently launched an initiative to give parents child-rearing advice. However, some commentators pointed out that Cameron was not the best parent himself: after all, he had once forgotten his child in a pub toilet. The argument ran: "David Cameron thinks there is too much bad parenting in Britain. But he once left his child in a pub toilet–so who is he to criticize?"

THE MISTAKE

Like all *ad hominem* arguments, this is simply *irrelevant*: the character of its proponent does not affect the truth of an argument. Even when the proponent himself engages in the behavior in question, it doesn't mean that he's wrong, or even insincere, in condemning that behavior. Just because the father smokes, this doesn't mean that he can't have an ubiased view that smoking is bad; still less does it show that he's *incorrect* when he says that smoking is bad.

THE COMEBACK

As with all *ad hominem* arguments, you should tell your opponent to stick to the argument, not worry about the person making it. After all, you are arguing about the facts of the matter, not about your respective behaviors. The father in the above example has a reasonable belief that smoking is unhealthy; the fact that he himself smokes does not in the least affect the truth of his claims.

If only things were that simple. After all, showing that your opponent can't reasonably hold a position based on contradictory commitments is a valid move (see p. 19). But a person's actions would seem to express just such implicit commitments (thus the old saying, "actions speak louder than words"): if I smack my children, I am, at least implicitly, committed to the view that it's *right* to smack my children. Hence, this argument can be a form of circumstantial *ad hominem*, directed at the commitments of the speaker; specifically, those to which his behavior commits him. The title of a book by G.A. Cohen puts it well: "If You're an Egalitarian, How Come You're So Rich?"

To respond to this sort of argument you have to show that you are not *committed* to your behaviors: that you do not engage in them because you think they are right, but because you are, in some way, weak. So, the smoking father could respond: "I don't smoke because I think it's good; I smoke because I'm addicted." Likewise, Cameron could respond: "I don't condone leaving children in pub toilets–I did so because I sometimes forget things" (of

course, whether such forgetfulness is a desirable trait in a states-man is a different matter).

The argument's name is Latin for "You too!" In other words, you say that this behavior is wrong; but *you too* engage in it. This is what the pot does when it calls the kettle black.

AFFIRMING THE CONSEQUENT

Formal	P is inferred from the major premise 'if P then Q' and the minor premise 'Q.'

Substantiating a statement by showing proof of a tangential consequence.

EXAMPLE

> "If I had a deadly disease, I'd have a cough. I have a cough. Therefore, I have a deadly disease."

REAL-LIFE EXAMPLE

Homer Simpson, worried about the (non-existent) problem of bears in Springfield, sets up a Bear Patrol. Looking around Springfield, he remarks with satisfaction: "Not a bear in sight. The Bear Patrol must be working like a charm!" His reasoning may be formalized thus:

Major Premise: *If the Bear Patrol were effective, then there'd be no bears around Springfield.*
Minor Premise: *There are no bears around Springfield.*
Conclusion: *The Bear Patrol is effective.*

The problem is that the institution of the Bear Patrol doesn't account for the true reason there aren't any bears around Springfield: that there weren't any to begin with.

THE MISTAKE

This argument is *formally* incorrect: that is, the conclusion simply *does not follow* from the premises. In the conditional 'if P then Q,' Q is a *necessary* condition for P: if P is the case, Q *must* be true. But P

is only a *sufficient*, not a *necessary* condition for Q: the antecedent doesn't *have* to be true just because the consequent is true.

So, in the above example, if I had a deadly disease, I'd have a cough. But it doesn't follow that if I have a cough I have a deadly disease.

Similarly with the *Simpsons* example, if the Bear Patrol *were* effective, there would be no bears in Springfield. But simply that there are no bears in Springfield doesn't show that the Bear Patrol is effective–Springfield just didn't have a bear problem in the first place.

THE COMEBACK

Since the fallacy is a formal one, the comeback is simply to point out the fallacy: that the speaker has violated a rule of logic. Still, the onus may be on you to show that he can't accept the converse conditional, 'If Q then P.' You can do so by showing that the converse consequence doesn't hold. So, suppose your hypochondriac friend says:

"I have a cough; but people with a deadly disease have a cough; so, I must have a deadly disease!"

He's implicitly committing this fallacy. You can reassure him by pointing out that, while everyone with a deadly disease may indeed have a cough, it's not the case that everyone who has a cough has a deadly disease–people have coughs for all sorts of relatively innocuous reasons.

In the *Simpsons* example above, Lisa tries to show that Homer's reasoning is wrong, using a parallel example: "This rock keeps away tigers. Look, it works: you don't see any tigers, do you?" Of course, Homer fails to recognize that *this* is fallacious too, and ends up buying the rock from Lisa.

Conditional statements let you make common but extremely useful inferences. So common and so useful are these inferences that they were given fancy Latin names by medieval scholars.

So, if I know that *if P then Q*, and I also know that *P* is the case, I also know that *Q* is the case: a form of inference known as *modus ponens* (the affirming way). Similarly, if I know that *if P then Q*, and I know that *not Q*, I also know that *not P*: if *P were* the case, then *Q* would have to be the case; but *Q isn't* the case, so neither can *P*. This is known as *modus tollens*: the denying way.

The fallacy of Affirming the Consequent is far less useful, but just as common, so it has its own Latin name: *modus morons* (the foolish way).

AMBIGUITY

Formal	An argument of the form "A is B, B is C, so A is C" (or similar), where the terms do not have a consistent meaning in the premises and conclusion.

An argument in which there is a term common to the premises and conclusion, or to more than one of the premises, but the term carries a different sense in each instance.

EXAMPLE

"Peter is a short professional basketball player. Hence, Peter is a professional basketball player, and he is short."

REAL-LIFE EXAMPLE

Plato's *Euthydemus* gives a truly wonderful example:

You say that you have a dog, Ctesippus.
Yes, a villain of one.
And he has puppies?
Yes, as bad as he is.
And the dog is their father?
Yes, I saw him mount the bitch myself.
And is he not yours?
To be sure he is.
Then he is a father, and he is yours; ergo, he is your father,
*   and the puppies are your brothers.*[2]

2   Plato, *Euthydemus*, translation adapted from Jowett.

If an argument is to be valid, its terms must have a consistent meaning. So, in a standard syllogism we say, "A is B, B is C, so A is C." But if the terms A, B, or C have a different meaning in each of the premises, then the syllogism just doesn't follow: the terms *look* alike, but have a different meaning each time they appear.

We can see this by working through the examples. The basketball example infers that "Peter is short" from "Peter is a short basketball player." But this equivocates between the *attributive* and *predicative* uses of the term "short." So, "Peter is a short basketball player" means something like, "Peter is below the average height of professional basketball players," whereas "Peter is short" means simply "Peter is of below-average height." But it's entirely possible for a short basketball player to be a tall man: if Peter is 6'5", he's below average for an NBA player, but is nonetheless a strapping young lad.

Plato's example, similarly, equivocates between two different senses of 'your *x*.' We can speak of Ctesippus's dog, or Ctesippus's father. But these denote very different relations: Ctesippus's dog is that animal to whom Ctesippus is related *as an owner*; Ctesippus's father is that person to whom Ctesippus is related *as a descendant*.

Now, there is a sense in which the dog *is* Ctesippus's father: the dog is a father, and Ctesippus owns a dog; the dog is thereby *Ctesippus's father* (compare: Ctesippus's dog is a mangy, flea-bitten mutt, so the dog is *Ctesippus*'s mangy, flea-bitten mutt). But clearly, this does not mean that Ctesippus's *relation* to the dog is that of son to father.[3]

---

3 NB: Aristotle's *De Sophisticis Elenchis* would analyse this as a *fallacy of accident.* Aristotle's examples of equivocation are barely translatable from the Greek, since they depend on specific characteristics of the Greek language, particularly its inflected nature. This is why it's a good idea for philosophers to learn other languages: inferences which sound natural in one language may become problematic when translated into another.

You need to be able to disambiguate the argument's terms, in order to show that the argument deploys them indiscriminately in different senses. Subsequently, you should restate the argument in terms that highlight the ambiguity. In the basketball example, you should recast the argument thus:

> P1. *Peter is below the average height of a basketball player.*
> ∴ *Peter is below average height.*

But now, the argument doesn't even *look* valid: it would presuppose the additional premise that basketball players are of average height, which is clearly false.

SIGNIFICANCE

Ambiguity embraces *equivocation*–a situation where the same word has different meanings. However, ambiguity is the wider notion. Equivocation, strictly speaking, is a semantic ambiguity: the ambiguity arises because the same word refers to different things (e.g., 'bank'). Ambiguity, on the other hand, can also arise because of the syntax of a sentence, endowing the same word with a different meaning in the context of different sentences. So, in the basketball example, "short" has a different meaning when qualified by the term "basketball player"; in the dog example, "your x" has a different meaning depending on what x refers to.

The development of logic was motivated, in part, by the need to develop a language immune to this sort of ambiguity. These languages are artificial–lots of squiggles and symbols replacing words–but are nonetheless extremely useful.

Some enthusiastic linguists tried to develop a syntactically unambiguous *natural* language. Lojban and Loglan are two examples of this; neither is extremely useful.

Syntactic ambiguity is frequently the source of humor, as illustrated by Groucho Marx's classic quip: "I shot an elephant in my pajamas. How he got into my pajamas, I'll never know," or the wartime newspaper headline: "French Push Bottles Up German Rear." But, as with *equivocations*, these belong in the music hall, not the debating chamber.

ANONYMOUS AUTHORITY

Informal	Argument *P* is justified by appeal to an authority A, whom the argument's proponent does not (or cannot) name.

An argument's proponent justifies it by appeal to an unidentified authority.

EXAMPLE

> *"Experts say that gluten is bad for you: hence, I should avoid gluten."*

REAL-LIFE EXAMPLE

The phrase "that's what they say!" or "experts say" is the most common form of this argument ("If you swallow gum, it wraps around your heart and you die!" "How do you know that?" "Well, that's what they say . . .").

THE MISTAKE

There's no problem with appealing to authority when that alleged authority is in fact an expert on the topic in question. So, if I say, "Black holes emit radiation," I can justify this by appealing to the authority of Stephen Hawking. But I cannot justify my belief when I cannot establish the credentials of the authority to which I appeal. So, it's illegitimate to say, "My mate Bob tells me that black holes emit radiation," if Bob knows nothing about physics. If we appeal to an *anonymous* authority, then their anonymity ensures that their credentials cannot be checked; hence, the appeal to this type of authority is illegitimate.

Challenge your opponent to establish the credentials of the authority he invokes. So, if your opponent says, *"They say* that swallowing gum is bad for you," ask him: "Who is this *they*, and where did *they* get their medical license?"

We meet with this kind of argument all the time–not just in official debating contexts, but in everyday discussions ("Eat your greens: everybody knows they're good for you!"). In an everyday context, of course, this is usually fine: when you brush your teeth every morning, you don't need to justify this by appealing to an expert. In a more formal debating context, however, such an implicit and unquestioning acceptance of authority just doesn't cut it: if you use an authority to back up your claim, you had better make sure of your authority's credentials.

APPEAL TO ANGER

Argumentum ad Odium

Informal	The proponent justifies his argument for *P* by playing on the anger of the audience.
	———————— OR ————————
	Proponent A argues *P*. Opponent B states that *P* offends him, therefore *P* must be false.

Attempting to defend a position by exploiting the audience's feelings of anger, bitterness and spite. Alternatively: attacking an opponent's argument on the grounds that it angers you or your audience.

EXAMPLE

> *"Let more immigrants into our country? These people who take our jobs, who live on welfare, who eat all kinds of foreign muck? I don't think so!"*

REAL-LIFE EXAMPLE

A lot of populist rhetoric falls under this category. Donald Trump is an especially instructive example: his wild denunciations of Washington, the media, immigrants, Muslims and other sundry figures, despite rendering him a figure of ridicule in the mainstream media, have been (at the time of this writing) instrumental in the success of his presidential campaign.

THE MISTAKE

Facts are facts, regardless of how one feels about them; just because a fact makes you angry, that doesn't stop it from being true. The

reverse also holds: if a demagogue whips up a crowd to believe that *P*, it doesn't make *P* true.

This is a difficult move to counter, not because the move is particularly complex, but because the opponent has, as it were, stopped playing the game of logic altogether: he's no longer trying to establish whether the argument is *correct*, but is either just cynically counting on mobilizing the audience's anger, or is himself too angry to reason properly.

So, although you could convincingly point out that your opponent is arguing fallaciously, this may be ineffective: logic, alas, has its limits. The better response would be to appeal to the audience's better nature: should your opponent try to whip up hatred against a certain group, respond in kind by stressing the need for tolerance and understanding (good luck with that).

SIGNIFICANCE

Logic is a powerful tool; its power, however, has its limits. So it frequently loses out against emotion, not because emotion is more reliable than reasoning, but because emotion is simply more *forceful*.

APPEAL TO AUTHORITY

Argumentum ad Verecundiam

Informal	Person A claims that *P*. A is considered an authority. Therefore, *P*.

Attempting to support an argument P, not by offering any direct evidence that P, but by appealing to the testimony of an authority A.

EXAMPLE

> *"My dad says that scientists just planted dinosaur bones to discredit creationism. Hence, evolution must be false!"*

REAL-LIFE EXAMPLE

Some figures of the early Christian church did not believe that the earth was round, because this contradicted the authority of the Bible. Thus, the Bishop of Gabala (4th Century AD) argued that the earth was flat, because this is what Scripture says: if it was round, how could the Lord "stretch out the heavens as a curtain, and spread them out as a tent to dwell in" (Isiah 40:22)? Fortunately, the views of these 'flat Earthers' never became official doctrine.

THE MISTAKE

There is no problem with justifying a belief by appeal to authority, provided that the authority is an expert on the subject at hand.

So if John is an expert on nuclear physics, it's reasonable to believe the claims he makes about, say, the possibility of cold fusion; however, it would be silly to defer to his authority qua nuclear physicist were he to pronounce on the death penalty.

THE COMEBACK

Demand proof of the authority's credentials on the subject under discussion. Does this person have expertise relevant to the topic at hand? If not, why on earth should you listen to them?

Even if this authority *does* have the relevant expertise, you can still raise doubts. An expert's opinion may not represent the consensus of other experts in the field; indeed, he may be in the minority, and other experts may treat him as a crank. Similarly, the expert may have a vested interest in getting people to accept his opinions: he may be misusing his expertise to gain a financial or other personal benefit. So, you could always demand a second opinion or corroboration from other experts in the field.

SIGNIFICANCE

We can't help but defer to authority in every stage and aspect of our lives. For instance, I may not understand aerodynamics but I have no problem boarding a plane, because I know it was designed by experts who assure me that it can fly. But we can still take such expert pronouncements with a grain of salt. After all, a group of 'experts' won't necessarily achieve an absolute consensus about a topic; experts aren't necessarily impartial; and they often pronounce on topics beyond their ken.

APPEAL TO CELEBRITY

Informal	Celebrity A believes that *P*. A is famous. Therefore, *P*.

Justifying a belief on the grounds that a celebrity believes it to be true.

EXAMPLE

"Noted actress Astrid McStar said that eating peas gives you cancer. So I'm going to stop eating peas!"

REAL-LIFE EXAMPLE

This argument is so silly that its advocates hardly ever spell it out. But most celebrity endorsements appeal, implicitly, to this kind of reasoning: Jennifer Aniston endorses L'Oréal shampoo, so I should buy L'Oréal! (After all, I'm worth it).

THE MISTAKE

This argument is a kind of *Argument from Authority* (see p. 43), but one in which the flaw is more readily apparent: being a celebrity almost never qualifies someone to pronounce authoritatively on any issue. That somebody had a Number One hit does not qualify them to pronounce on science; we should not listen to someone's political opinions just because their latest movie was a success.

THE COMEBACK

As with any appeal to authority, the proper response is to demand appropriate credentials. Jennifer Aniston may say L'Oréal shampoo is the best, but she's not a trichologist.

There are, of course, exceptions to this rule. If a sportsman endorses a certain brand of trainers, his testimony has some credibility: after all, he needs good footwear to succeed in his game. Similarly, it might be the case that Jennifer Aniston has such fabulous hair just because she washes it with L'Oréal every night (though I'm inclined to skepticism on that one). Of course, we may still wonder whether the celebrity is objective in his endorsement: after all, he earns a fee for it.

SIGNIFICANCE

People fall for this fallacy unconsciously more than consciously: they may believe things because celebrities say so, but they rarely *justify* their beliefs in this way (when the argument is spelled out, it is patently absurd). Still, the weight of celebrity is very significant: why else would advertisers seek celebrities to endorse their products?

APPEAL TO COMMON BELIEF

Argumentum ad Populum

Informal	Everybody believes that *P*. Therefore, *P*.

Justifying a proposition on the grounds that many people suppose it to be true.

EXAMPLE

> *"Everybody knows the sun goes 'round the earth; therefore, the sun goes 'round the earth."*

REAL-LIFE EXAMPLE

Urban myths provide an excellent example of this phenomenon. "Don't flush your baby crocodile down the toilet–everybody knows it will grow up and come out to bite people on the butt!"

THE MISTAKE

This kind of argument requires a suppressed premise: "If everybody believes a certain something, it must be true." But this principle is false. People are fallible, be they individuals or groups. As the old song goes, "they all laughed at Christopher Columbus/ when he said the world was round." Hence, we shouldn't automatically give credence to a belief just because many or most people believe it.

THE COMEBACK

The best comeback is to show that there is good evidence that the majority is wrong. For instance, you could invoke an expert's opinion asserting that the popular belief is false; even better, you could offer evidence that contradicts the majority's view.

This is another example of fallacies that are, for the most part, perpetrated unintentionally: cases where people just don't think enough about their beliefs, but rather espouse them unthinkingly just because other people do.

Now there have been those who defend this practice by invoking the so-called 'wisdom of crowds': whereas an individual will by definition have limited experience and be prone to making mistakes, such individual errors will tend to 'level out' when there are enough people pitching in.

But we should also remember that the 'wisdom of crowds' tends to allow false beliefs to become common sense: if everybody believes something just because everybody else believes it, then it's entirely possible that they all believe something false. This is particularly notable in prejudicial beliefs: people in a racist group may unquestioningly believe that members of a different race are inferior just because everybody else around them believes this. The 'wisdom of crowds' is often a byword for collective folly.

Incidentally, the line about Columbus quoted above is *itself* an example of the *ad populum* fallacy. Nobody laughed at Christopher Columbus because he said the world was round: people had known that since the time of Ancient Greece. They laughed at Columbus because he'd calculated the earth's size erroneously, which led him to claim that it was far smaller than it actually is. On the basis of this miscalculation, he set out seeking a Westward passage to India; it was a serendipitous stroke of luck that there happened to be a huge, undiscovered continent in the way.

APPEAL TO DESPERATION

The Politician's Syllogism

Informal (although it's also been expressed formally)	Situation S demands a response. Action *P* is proposed as a solution, where *P* is, in fact, irrelevant to S.

Demanding that an action be performed to resolve a situation, regardless of whether the proposed action will in fact resolve the situation in question.

EXAMPLE

> This type of argument was canonically formulated in the British sitcom **Yes, Minister:**
>
> P1. *Something must be done!*
> P2. *This is something.*
> ∴ *We must do this!*

REAL-LIFE EXAMPLE

In 2012, the Republicans proposed raising the age of Medicare eligibility, in an effort to stem the rising budget deficit; research showed, however, that doing this would in fact *increase* the deficit in the long run. Paul Krugman, commenting, analyzed the Republican proposal thus:

> P1. *Medicare costs demand a serious response.*
> P2. *Raising the Medicare age eligibility sounds like a serious response.*
> ∴ *Therefore, the Medicare age eligibility must be raised!*

THE MISTAKE

We can analyze the problem with this fallacy in a number of different ways. It can be analyzed as a *formal* fallacy, namely an *Equivocation* (see p. 108), in the sense that the 'something' doesn't mean the same thing in *P1* and *P2*. In *P1*, the word 'something' is shorthand for "something *that will resolve the crisis* must be done"; in *P2*, however, it lacks this qualification. Hence, the conclusion doesn't logically follow.

Alternatively, even if the proposed solution would resolve the problem at hand, it doesn't mean it should be implemented; there may be other, possibly better ways of resolving the problem. Moreover, the proposed solution may be worse than the original problem. So, if I had an infected toe, one solution would be to amputate this body part. However, this doesn't mean I *should* amputate my toe–I could, for instance, treat it with antibiotics; or I might opt to just bear the infection awhile.

THE COMEBACK

You can point out that the proposed solution will prove ineffective; that there are alternative solutions to the problem at hand; or that the proposed cure is worse than the disease. Best of all, use all three comebacks (provided that they don't contradict each other, of course!).

So, if someone argues: "Marijuana use is a problem. To solve it, we should lock up teens who smoke pot!" you can respond that imprisoning drug-users is *ineffective*, because it doesn't stop people from using drugs; that there are better ways of dealing with this problem (e.g. drug education, harm-reduction programs); and that imprisoning drug-users in fact creates more problems than it solves (you're generally worse off going to prison than smoking pot).

This fallacy is (thanks to its appearance in the British sitcom *Yes, Minister*) sometimes known as the *Politician's Syllogism*. There, a senior civil servant compares this type of inference to the syllogism:

P1. *All cats have four legs.*
P2. *My dog has four legs.*
∴ *Therefore, my dog is a cat.*

This fallacy, then, could be taken as an example of *Affirming the Consequent* (see p. 29). The first premise says that there is a course of action that has to be implemented. The second premise says that *this* is an action. However, it doesn't follow that this *particular* action *has* to be implemented: for that, we would need an additional premise stating that *any* action must be implemented.

APPEAL TO EMOTION

Informal	Proponent A argues for or against conclusion P by invoking the emotional effects of P.

Arguing for the conclusion of an argument by appealing to the emotions of the audience, rather than addressing the matter at hand.

EXAMPLE

"Reducing welfare payments is cruel. Hence, we should not reduce welfare payments!"

REAL-LIFE EXAMPLE

Any time you hear the phrase 'think of the children!' or a variation on it, you can be sure that this fallacy is lurking. Advocates of California's Proposition 8, for instance, or the *Protecting Children from Internet Pornographers Act* of 2011 (which required Internet Service Providers to retain user IP addresses, and enable the government to access that information on demand) would make appeals of this nature. The effect of this rhetoric was to make a rational response impossible: "you oppose this bill? What sort of *monster* are you?!"

THE MISTAKE

There is no problem with appealing to emotion in pursuit of a *pragmatic* end (i.e., to motivate your audience to do something)–we frequently need that tug on our heartstrings to goad us into action. The mistake arises when the appeal to emotion is used *in lieu* of an argument. The facts of the matter may be frightening, disgusting, enraging: but they are still the facts, regardless of how one feels.

Arguing against this move can be risky: if your opponent has successfully mobilized the audience's emotions, that audience might be prejudiced against considering the facts at all. Under these circumstances, even if you simply put forward the facts of the matter in a rational and objective manner, you run the risk of appearing cold-hearted and callous. The better response may be to buttress your own arguments with alternative appeals to emotion. So, if you can argue that your opponent's course of action would indeed stop *some* suffering, but your own course of action will prevent *even more* suffering, then you can make an emotional appeal yourself: doesn't your opponent care about the preventable suffering of *these additional* people?

SIGNIFICANCE

This fallacy is the generic form of which other appeals to emotion—the *Appeal to Fear* (see p. 58), the *Appeal to Pity* (see p. 73)—are subspecies. The fallacy is identical in all cases: the proponent tries to argue his case not by appeal to the facts, but on the strength of his and others' emotional reactions to the facts. It is a fallacy for the very same reason: my feelings about the facts of the matter are simply *irrelevant* when it comes to their truth or falsity.

This fallacy also occurs when what's at issue is not a factual matter per se, but instead turns on a question of right and wrong. So, an opponent of homosexuality may say: "Homosexuality is wrong: it's *disgusting.*" The problem here, of course, is that his emotional reaction to homosexuality has no bearing in demonstrating that homosexuality is either right or wrong. After all, different people have different attitudes to homosexuality; why should *his* emotional reaction have more weight than anyone else's?

APPEAL TO FAITH

Informal	Proponent A has faith that P. Therefore, P.

Arguing for a conclusion purely on the basis of faith, rather than invoking any reason or evidence for its truth.

EXAMPLE

"I have faith that the Lord will protect me from cancer. Hence, I have no need to give up smoking."

REAL-LIFE EXAMPLE

Almost any religious dogma will, almost by definition, rely on this fallacy to some extent. In defending these dogmas, some apologists will also appeal to reason: St. Thomas Aquinas's *Summa Theologica*, for instance, is a masterpiece of subtle and profound argumentation in defense of Catholic dogma. Other religious apologists relegate a much greater role to pure faith. For instance, here is the inimitable Tertullian:

> What has Jerusalem to do with Athens, the Church with the Academy, the Christian with the heretic? . . . I have no use for a Stoic or a Platonic or a dialectical Christianity. After Jesus Christ we have no need of speculation, after the Gospel no need of research. . . . Since finding was the object of your search and belief of your finding, your acceptance of the faith debars any prolongation of seeking and finding.[4]

---

4 Tertullian, *Prescriptions against the Heretics*

Tertullian goes on in this vein, but his message is already abundantly clear: reason has no place in justifying dogma–faith alone can justify it.

This type of argument simply gives no *reason* to accept the truth of its conclusion: its proponent substitutes evidence or reasoning with a blanket appeal to faith. But cogent arguments require reasoning and logic; an appeal to faith furnishes neither.

Furthermore, such appeals are ineffective, precisely because faith is not universal. Faith is not of one kind, but many; moreover, many people lack faith altogether. A proponent's appeal to faith only works for other people who share that faith; it will be utterly ineffective against those who don't.

You can simply point out that, by appealing to faith, your opponent has pretty much conceded that he has no valid reasons for his argument. But this move may be ineffective, depending on the strength of his faith and that of your audience. It may be better, then, to counter him on his own turf, either by showing how an appeal to faith can justify *anything*, or by using his faith to support *your* conclusion.

So, if your opponent uses Leviticus's proscription on homosexuality to oppose gay marriage, you could argue: "Leviticus *also* prescribes that you shouldn't wear mixed fibers: but you are wearing a polyester suit!" Alternatively, you could say: "Fine, that's the faith of the Old Testament, but what about the message of tolerance and love in the New Testament? Shouldn't that encourage us to accept homosexuality?"

This fallacy is typically an *Appeal to Authority* (see p. 43): I believe something because this holy book, or this religious official, or the light of the Holy Spirit tells me to believe it. These arguments are problematic, because there are lots of holy books and lots of religious officials, and they don't all agree on what to believe. For instance, Islam teaches that God is a single entity, while the Catholic Church claims that He is triune; other religions may teach that there are many gods. But if the appeal to faith can justify such radically different conclusions, why should we believe *any* of them?

APPEAL TO FEAR

Argumentum ad Metum

Informal	Either *P* or *Q*. *Q* is frightening. Therefore, *P*.
	──────── OR ────────
	P is presented in a way that plays on the audience's preexisting fears.

Justifying a conclusion by instilling fear against the alternatives in your audience. Alternatively: justifying a course of action by playing on the audience's fears.

EXAMPLE

> *"Either there's a monster under my bed, or there isn't. But I would be terrified if there were such a monster! So, best to assume that there isn't."*

REAL-LIFE EXAMPLE

This type of argument is common enough among fire-and-brimstone preachers, who tell you that if you don't believe in God you will burn in the fires of hell! In other words, you should believe in God because you should fear the consequences of *not* believing in God.

For the second kind of *argument from fear*, take Donald Trump's election proposal to ban Muslims from the United States. Trump gave very little justification for his proposal, or the proposal's presumed effectiveness (or, indeed, its constitutional legitimacy); rather, his argument played on the audience's fear of Muslims, with demagogic language that demonized them: "Something bad is happening . . . Something really dangerous is going on."

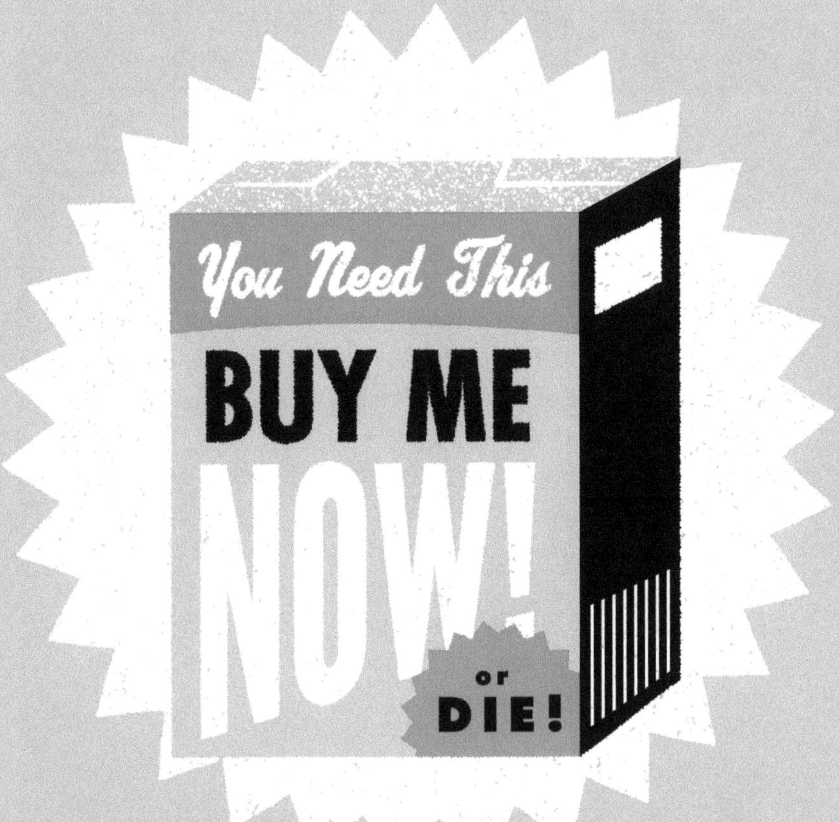
You Need This
BUY ME
NOW!
or
DIE!

As with any *Appeal to Emotion* (see p. 53), this argument is simply *irrelevant*: the facts are the facts, regardless of how I feel towards them. Whether I am frightened of a conclusion or not, that doesn't stop it from being true.

This argument is more frequently used in contexts concerning *actions*. So, in a case where action P or Q must be taken, the proponent argues for P by making Q look scary; alternatively, he justifies an action by playing on the audience's fears. This is not, as such, invalid: Churchill's arguments for war were no less valid because they played on fears of Nazi Germany. But such appeals are invalid when the relevant fears are ungrounded.

THE COMEBACK

The trick here is to show that there's nothing to be frightened of, or that your opponent is exaggerating. So, if I don't believe in God, I don't believe in hell anyway; so why worry?

To make this comeback really effective, however, you have to be in command of the facts. So, if your opponent is arguing that the US should invade a small country on the grounds that it has stockpiled a terrifying arsenal of weapons, ask him to prove this latter claim. If he can't, you can demonstrate that he is just playing on his audience's fears. Whether this comeback will change anybody's mind, however, is another matter.

SIGNIFICANCE

This argument is is, sadly, far too popular among demagogues, who try to shore up their power by playing on their followers' fears of their opponents or minority groups. Whether it is a case of a military junta playing to fears of their socialist opponents, or an anti-Semitic government playing to a fear of Jewish people, the principle remains the same: play to the tune of your audience's fears to make one group look like their enemy and to make yourself look like their allies.

More prosaically, this is also a favored tactic of advertisers. "Buy deodorant," they will say; otherwise, you will stink! "Baby-proof your house (with our products): otherwise, your baby will drink your bleach!" Spotting such fallacies doesn't just help with your logic, but with your shopping as well!

APPEAL TO HEAVEN

Informal	"God demands that *P* must be done. Therefore, *P* must be done!"

Justifying an action on the grounds that it has divine assent, in other words, that God wants you to engage in it.

EXAMPLE

"Why did I rob all those banks? Because God told me to!"

REAL-LIFE EXAMPLE

The Biblical story of Abraham and Isaac (Genesis 22:2-8) perfectly illustrates this fallacy. God commands Abraham to sacrifice his son Isaac on an altar. Abraham, being a pious man, follows God's instructions, and binds Isaac to an altar, fully prepared to sacrifice him. Abraham is only stopped at the last minute by an angel, who, saying "now I know you fear God," allows Abraham to sacrifice a ram in Isaac's stead.

THE MISTAKE

This argument is a kind of *Appeal to Faith* (see p. 55), and consequently suffers from the same problem: even assuming that there *is* a God, it can be tricky to determine His will. Your opponent may believe that his actions are divinely mandated; however, the burden of proof is on him to show that he's justified in assuming this divine warrant. After all, God has been used to justify all sorts of heinous acts, be they serial murders or widespread religious persecution.

Your opponent had better have some compelling evidence to show that God really demands what he's proposing.

THE COMEBACK

You can simply dismiss this kind of argument by showing it's an appeal to faith: your opponent may claim that God demands that such and such be done, but nobody else has reason to believe this.

Alternatively, you could ask *why* God demands this. Does God command it because it is good? If so, the *Appeal to Heaven* is superfluous: the action is justified anyway, because it is good. Or is it good just because God commands it? But if God's will makes actions right or wrong, then this could justify *anything*, even killing your first-born son (as the Abraham and Isaac example shows). Can we really justify actions on the say-so of one individual, no matter how powerful He may be?

SIGNIFICANCE

The latter response to the *Appeal to Heaven* is sometimes called the *Euthyphro Dilemma*, based on an analogous argument in Plato's *Euthyphro*. Euthyphro, a religious fanatic, believes he has divine warrant to prosecute his father for murder: his action is holy; it is loved by the gods. Socrates asks him, simply: "Do the gods love something because it is holy, or is it holy because the gods love it?" Euthyphro lives his life according to (what he takes to be) divine edict, yet he is at a loss to answer even this simple question about why divine edict is valuable.

APPEAL TO THE MOON

Informal	Society S, or person P, has accomplished feat F. Therefore, society T, or person Q, should be able to achieve feat G!

Arguing that, because a person or society has achieved something great (for example, putting a man on the moon), another person or society should be able to achieve something else of similar stature.

EXAMPLE

> *"If we can put a man on the moon, we can sure cure cancer!"*

REAL-LIFE EXAMPLE

After the 2008 financial crisis, the government spent hundreds of billions to ensure that certain major banks and other financial institutions didn't go under. Hence, there was less money to spend on social welfare, defense, infrastructure, etc. But, critics argued, if the government could afford to spend billions on bailing out the banks, surely it could afford a similar sum to fund social welfare programs?

THE MISTAKE

We can understand this type of argument as a sort of *Appeal To Possibility* (see p. 75). So, the reasoning goes, society achieved the initial great feat, which was considered impossible; hence, it can certainly achieve this second thing, even though it is likewise considered impossible. But the two feats are different. Therefore, it doesn't follow that it's even *possible* to achieve the second feat, let alone that it's *probable*.

THE COMEBACK

First, point out that your opponent's argument is simply *invalid*: the fact that one difficult thing has been achieved doesn't mean that a different difficult thing may *also* be achieved. After all, the difficulties associated with the latter feat remain unaffected by the achievement of the first feat. You should then point out just how great these difficulties are; perhaps putting a man on the moon is in fact relatively simple compared to curing cancer.

SIGNIFICANCE

This argument may be formally invalid, but, when used properly, its underlying message seems sound. Don't be scared about dreaming big: people can, and do, achieve great things. After all, we put a man on the moon!

APPEAL TO NATURE

Informal	*P* is natural, therefore *P* is good; or, *P* is unnatural, therefore *P* is bad; or, *P* is natural, *Q* is unnatural, therefore *P* is better than *Q*.

Grounding the value of something by appealing to its naturalness; in other words, claiming either that something is good because it is natural, or bad because it is unnatural.

EXAMPLE

> *"Homosexuality is unnatural: the body just isn't designed for that sort of activity. So it must be wrong!"*

REAL-LIFE EXAMPLE

The following advertisement for American Spirit neatly encapsulates the *Appeal to Nature*:

> *TASTE NATURE.*
> *AND NOTHING ELSE.*

> *You'll never find any additives in our tobacco. What you see is what you get. Simply 100% whole-leaf natural tobacco. True authentic tobacco taste. It's only natural.*

The ad implicitly assumes (or invites the reader to assume) that, while the nasty "additives" are bad, the tobacco itself must be wholesome, because it's natural. They forget to mention, of course, that this "all natural tobacco" is precisely what kills you.

THE MISTAKE

This type of argument assumes, implicitly or explicitly, that what's 'natural' is good, while what's 'unnatural' is bad. But this dichotomy is obviously false: jam is 'unnatural,' death cap mushrooms are 'natural,' but that doesn't mean that death cap mushrooms are better than jam on toast.

THE COMEBACK

Your comeback should be tailored to your opponent's argument. Sometimes, your best recourse would be to challenge him on the implicit dichotomy 'natural=good,' 'non-natural=bad.' So, if your opponent argues that, say, herbal medicine is better than pharmaceutical medicine, on the grounds that herbal medicine is 'natural,' you should point out that pharmaceutical medicine is usually rather more effective at fighting disease.

Alternatively, you could question whether your opponent's distinction between 'natural' and 'non-natural' makes sense. So, if your opponent argues that homosexuality is wrong because it's unnatural, you could argue that homosexual behavior is also observed in animals—how, then, could it be deemed unnatural? Further, you might question whether we should be making this distinction at all. Isn't man part of nature? Isn't his intelligence and inventiveness part of his nature? How, then, can anyone claim that the products of his intellect are somehow 'unnatural'?

SIGNIFICANCE

This argument has been often used by quacks of all stripes, be it the proponents of 'natural' medicines, Social Darwinists (and their charming National Socialist progeny), or traditionalist moralists. All of them set up a *normative* dichotomy between the 'natural' and the 'non-natural'; a dichotomy which is, as we've seen, not unproblematic.

The distinction between 'natural' and 'non-natural' (or 'artificial' or 'conventional') in Western philosophy starts with the Sophists (c. 5th-4th centuries BC), some of whom used this distinction to denigrate the conventional in favor of the natural. One particularly memorable example of this is furnished by Callicles, an otherwise unknown figure making an appearance in Plato's *Gorgias*. In that dialogue, Callicles denigrates 'conventional' (democratic, egalitarian) morality in favor of the doctrine of 'might makes right,' on the grounds that it is the universal law of nature that the strong should rule the weak. Against this claim, Socrates delivers a knockdown argument: if the many are *collectively* stronger than any one individual, no matter how strong he may be, then they have (by Callicles's own principles!) the right to rule the one strong man.

NB: this argument is sometimes conflated with the *Naturalistic Fallacy* (see p. 150). But they are very different: the *Naturalistic Fallacy* concerns what sort of property goodness is, not whether nature is 'good' or not.

APPEAL TO NORMALITY

Informal	P is normal, therefore P is good; alternatively, P is abnormal, therefore P is bad.

Judging whether something is good or bad depending on whether it is determined to be normal.

EXAMPLE

"Normal *people listen to Top 40 hits, not to Bach. So, listening to the* Brandenburg Concertos *is wrong!*"

REAL-LIFE EXAMPLE

President Nixon's appeal to the "Silent Majority" has echoes of this fallacy. In response to the political turmoil of the late '60s, Nixon contended that the protests against Vietnam, or in favor of civil rights for women and minorities, no matter how vocal, did not represent the views of 'ordinary' Americans. This rhetoric encouraged people to ignore the protesters on the grounds that they were 'abnormal.' The underlying logic seemed to be: "*You* are the ordinary, normal Americans; *they* are the 'other,' the minority; you should ignore them, because they are abnormal."

THE MISTAKE

Sometimes, what's 'normal' is legitimately a byword for what's 'good'; for instance, if a doctor says "he's breathing normally," that means the patient is fine; if the doctor says "he's breathing abnormally," there's usually a problem. But when what's 'normal' is just what most people do or believe or like, then it doesn't have this *normative* connotation. Just because a majority of people prefer Justin Bieber to Bach, it doesn't mean the former is in any way superior to the latter. Additionally, the opinions of the majority can shift, as in the case of Nixon's appeal to the then "silent majority"; thus, some views of the then 'minority' have now become 'normal': who nowadays would defend Nixon's values?

THE COMEBACK

You can point out that, in matters of subjective taste, the majority's opinion doesn't make things right or wrong; just because what you like may be unpopular, you aren't *wrong* in liking it.

Moreover, by valuing what's normal, and denigrating what's different, your opponent is committed to a rather uninspired and conformist view of the world, one in which there's no room left for individuality or non-conformity. But the world needs people who don't go along with the majority opinion: they are the visionaries who blaze new trails in the arts, letters, and sciences.

SIGNIFICANCE

This argument is a form of *Argumentum ad Populum* (see p. 47): if the *majority* of people (i.e., the 'normal' people) think something is right, then they *must be* right. But this inference is not only invalid but dangerous, laying the ground for what J.S. Mill called the *tyranny of the majority*: a state of affairs where minority opinions could be considered wrong or evil just because they go against the majority. True, a healthy society needs conformists, but it equally needs people who go against the grain and follow their own path.

APPEAL TO PITY

Argumentum ad Misericordiam, or The Galileo Argument

Informal	Argument *P* is justified by invoking the opponent's pity.

Attempting to support a position not by offering any arguments or evidence in its favor, but by appealing to the opponent's feelings of pity or guilt.

EXAMPLE

> *"I know I got every question in the exam wrong, but I need an A to get a scholarship. Hence, you should give me an A!"*

REAL-LIFE EXAMPLE

An old joke puts it well: "I know you've caught me killing my parents with an axe. But I don't deserve to be punished! Can't you see I'm an orphan?"

THE MISTAKE

This fallacy is a specific kind of *Appeal to Emotion* (see p. 53), and shares the common problem plaguing these types of fallacy: one's feelings are *irrelevant* to the facts. If I argue that two plus two makes five, it *doesn't matter* that I may be dying, just got divorced, or lost a limb; pity me or don't, I'm still wrong.

THE COMEBACK

The simplest comeback would be to point out the argument's irrelevance: your opponent isn't supporting his position at all, but simply tugging at the heartstrings. Of course, you might look heartless in doing so. As an alternative, you could turn his rhetorical technique against him, by pointing out how cynical he's being in using this ploy: he's just telling the audience a sob story to conceal his faulty or nonexistent argument. By exposing him thus, you could get him to lose the good will he's trying to attract.

SIGNIFICANCE

This argument is very popular in legal defenses centering on the notion that one should exonerate the criminal because of his pitiable circumstances. In some cases, such a defense is valid–you don't prosecute a man who steals a loaf of bread to feed his starving family. In other cases it's misplaced, as with the orphan example above.

This argument is sometimes called the *Galileo Argument*, named after the sufferings of the great Italian scientist. This is a bit of a misnomer; Galileo's arguments were based on rigorous observation and logical deduction, not pity. In fact, it was his watertight logical arguments that brought him into his eventual pitiable state.

APPEAL TO POSSIBILITY

Informal	*P* is possible, therefore *P*.

Asserting that something is *or* will be *the case on the grounds that it's* possible *that it is the case.*

EXAMPLE

> *"It's* possible *that the sun won't rise tomorrow.*
> *Hence, the sun isn't going to rise tomorrow!"*

REAL-LIFE EXAMPLE

'Catastrophizing'–a mode of thinking where somebody tends to assume that the worst-case scenario is going to happen–is an excellent example of this. You imagine the worst thing that can possibly happen to you, and then convince yourself that it's going to happen–you'll lose your job, your wife, your kids, you'll end up homeless, etc., even though you have no evidence that *any* of this will happen.

Faulty as this thinking is, it has nonetheless been immortalized in Murphy's Law: "If anything *can* go wrong, it *will* go wrong"; to which Sod's Law adds the addendum: ". . . and at the *worst possible time.*"

THE MISTAKE

The argument is, at its simplest, a *modal* fallacy: not everything that is possible is *actual*. After all, it's *possible* that Mars is inhabited by intelligent turnips, or that the Soviet Union could have won

the Cold War, or that the sun won't rise tomorrow; however, that doesn't make any of these statements true.

Things become more complex when we're dealing with *probabilities*: when the envisaged outcome will *probably* but not *certainly* happen. So, it's *probable*, let's say, that Too Good for Glue will win the Kentucky Derby; his odds are at 1:4. But it's not *certain* that he will win. You would probably be better off not betting your life savings on it.

THE COMEBACK

Your opponent tries to get you to conclude that something *is* the case because it *could* be the case. But there's a pretty big gap between *possibility* and *actuality*. You can ask how *probable* it is that what he's saying is, or will come, true. So, suppose he believes the moon landings were faked: after all, isn't this *possible*? But the moon landings involved thousands of people; if they were faked, this would require that *all* of them somehow kept this a secret, rather than even *one* of them earning a fortune by selling the story to the press. Just how likely is that scenario?

SIGNIFICANCE

This fallacy seems to underlie others, such as the *Slippery Slope* (see p. 175) fallacy. So, it's certainly *possible* that a minor change will lead to another, and then to another, and so on, until you have a truly catastrophic change. But, you may respond, it doesn't mean that any of this *will* happen, or even that it's *likely* that it will happen.

APPEAL TO RIDICULE

Reductio ad Ridiculum

Informal	Proponent A argues that *P*. Opponent B undermines *P* by ridiculing it, without addressing the argument underpinning *P*.

Attacking an opponent's argument not by addressing the matter at hand, but by resorting to mockery: for example, repeating his argument in a sarcastic tone.

EXAMPLE

> *"So you say God, this* magic sky pixie, *made the world. Please tell us more about this* magic sky pixie, *I'd love to hear it!"*

REAL-LIFE EXAMPLE

During the 2008 presidential election, Obama claimed that we could all do our bit to reduce America's dependence on oil; for instance, if you inflate your car's tires properly, it saves on gas. His opponent John McCain and his various allies in the conservative media had a field day with this. So, here is Senator McCain's response:

> *"He actually thinks that raising taxes on oil is going to bring down the price at the pump. He's claiming that putting air in your tires is the equivalent–is the equivalent of new offshore drilling. That's not an energy plan, my friends. That's a public service announcement!"*

There's no real argument here, of course: Senator McCain just lays out his opponent's position and tries to make it sound as silly as possible.

THE MISTAKE

This just *isn't an argument*. You opponent isn't marshaling any new evidence, or showing why your argument is invalid, or adding any new substantive content to the debate: they're simply making your argument sound silly.

THE COMEBACK

Ask your opponent whether he actually has a rebuttal, or whether he just wants to waste the audience's time putting on a silly voice.

SIGNIFICANCE

This type of argument may not be valid, but can nonetheless be effective. By mocking an opponent's argument, they're not only implying that their *argument* is foolish, but suggesting that the *opponent himself* is foolish for espousing it. But, crucially, they're also getting their *audience* to believe that *they* would be foolish for believing the argument: "Only stupid people would believe my opponent's argument—and *you*'re not stupid . . . are you?" The *Appeal To Ridicule* effectively shuts down critical scrutiny of your position, and that's why it can be so effective.

APPEAL TO TRADITION

Informal	P is traditionally believed to be true. Therefore, P. (Implicit premise: whatever has been traditionally believed to be true is true).

Arguing that something is true, or valuable, on the grounds that it is traditionally believed.

EXAMPLE

"A woman's place is in the home; after all, this has traditionally been the case since time immemorial!"

REAL-LIFE EXAMPLE

American football has come under scrutiny recently because of a growing awareness of the grave long-term cranial and spinal damage its players suffer. In response to calls to ban the sport from high schools and universities, some defenders of the sport made an *Appeal To Tradition*: of course we should keep the sport the way it's always been, it's a time-honored American tradition, they claimed. These commentators did not address why a sport that causes such injuries is a worthy tradition.

THE MISTAKE

Your opponent argues, at least implicitly, that what's traditional is good *because* it's traditional. But clearly, there are traditions that are wicked (the practice of female genital mutilation, for example), as well as traditions that are mistaken (the once-popular belief that the sun goes round the earth, for instance). But if traditions can be

debunked, then an appeal to tradition cannot do the work of showing that something is good or correct.

THE COMEBACK

Begin by pointing out that your opponent's premise is unsound: tradition alone can justify nothing. Then, you can point out just how harmful your opponent's tradition actually is. So, a woman's place may *traditionally* have been in the home, but this was hugely detrimental to their welfare; society may *traditionally* have thought homosexuality was sinful, but this was an injury to homosexuals; people may have *traditionally* smoked in bars, but we now know this was harmful to the health of staff and patrons.

SIGNIFICANCE

This mistake is a type of *Argumentum ad Populum*; however, the *populum* to whom the *argumentum* is *ad* comprise the dead as much as the living: we should believe something, because the majority of our ancestors believed it. But, as with the *Argumentum ad Populum*, what our ancestors believed wasn't necessarily right; indeed, their ignorance was compounded by *their* faith to tradition.

Bear in mind, however, that traditions can encode societal wisdom, even if that wisdom is not immediately apparent to its practitioners. The traditional strictures against sex outside of marriage, for instance, form part of a network of beliefs and practices about marriage, childbearing and legitimacy; to attack these strictures risks undermining all of these other practices, on which societal stability depends. Of course, the burden of proof is on the person defending these practices.

ARGUMENT FROM IGNORANCE

Argumentum ad Ignorantiam

Informal	Proponent A argues that *P*, on the grounds that there is no evidence that *P* is false; alternatively, he argues that *P* is false on the grounds that there is no evidence for *P*.

Justifying a conclusion by appealing to the lack *of evidence that it is false; alternatively, assuming that something is false because of* lack *of evidence that it is* true.

EXAMPLE

> *"Why don't you believe that aliens walk among us as we speak? What evidence do you have to suggest this view is wrong?"*

REAL-LIFE EXAMPLE

In 1950s America, Senator Joseph McCarthy conducted a notorious witch-hunt against communists and communist sympathizers, whom he believed to have infiltrated American institutions at the highest level. Giving evidence against alleged communist fellow travellers in the State Department, Senator McCarthy said of one case:

> *"I do not have much information on this except the general statement of the agency that there is nothing in the files to disprove his Communist connections."*

In other words, we should assume that the agent in question was a communist, because we don't know that he is *not* a communist.

Just because we don't know something, or can't readily find evidence in its favor, that doesn't necessarily mean it's false. Of course, in most cases, it's a good idea *not to believe* something if you don't have adequate evidence *for* it. But even if I should *not believe* that something is true, it doesn't follow that I should *believe that it is false*: agnosticism pending further evidence is always an option. Certainly, if I should *not believe* that something is false, having no evidence for its falsity, this doesn't mean that I *should* believe it to be *true*.

Remind your opponent of the crucial distinction between *absence of evidence* and *evidence of absence*. So, just because there is no positive evidence for something, that is not, by itself, a reason to *dis*believe it. There may be other reasons that the evidence is not forthcoming. Thus, we may have no evidence of aliens, but that doesn't mean we should disbelieve their existence: the universe is a big place, and they may be hiding somewhere.

These considerations of the available evidence are especially important when this argument is deployed to prove a negative: if you can't prove me wrong, I'm right! First, remind your opponent that *he* needs some positive proof that he's right. Second, point out that he's essentially asking you to prove a negative, which is rather difficult. After all, how could you *disprove* the existence of aliens, short of exploring every inch of the universe? If he doesn't budge, you can reduce his position to absurdity, by showing that his move lets you justify *anything*. So, if you accept his argument, he should also accept that there are fairies at the bottom of the garden; after all, he can't prove otherwise and, by his logic, if you can't disprove something you must accept it.

This argument might be understood as resting on a *false dilemma*. So, either I should believe *P*, or not; if I have no reason to believe that *P*, the only alternative is to believe that *not-P*. But the negation of 'believing P' is not '*dis*believing P,' but simply '*not believing* P,' a condition satisfied by the agnostic as much as the atheist.

This type of argument, as we've seen, sometimes involves *Proving Non-Existence* (see p. 159): if I can't show that *P* is false, I must believe that *P*. But *Proving Non-Existence* is a tricky task.

BASE RATE

Informal	In determining the probability of an event E, the *base rate* probability that E will happen is disregarded, and specific facts about the case are used instead.

Information about the overall probability of an event is ignored when estimating how likely it is to occur in a particular case.

EXAMPLE

Suppose you have developed a test for disease X, which, although rare–only one person in a thousand suffers from it–is always fatal. Now, your test gives no false negatives, but has a 5% chance of giving a false positive, meaning that if you have the disease, the test will always tell you; but even if you don't have the disease, you will get a positive result 5% of the time.

Now, suppose Bob takes the test for the disease, and–oh no!–his result comes up positive. Of course, it may *be a false positive–but only 5% of cases are false positives. So, Bob thinks, there's a 95% chance he's got the disease; it's time to put his affairs in order.*

Not so fast. Take a thousand randomly selected people from the population, and run the test on them. Of this thousand, one person will have the disease; he will display a positive result. But, as we've said, the test has a 5% false positive rate. Hence, there will be 50 additional people who test positive for

disease X. So, the probability that Bob has disease X
is in fact 1 in 51: the overwhelming likelihood is that
he doesn't *have the disease.*

Practitioners of faith healing and 'alternative' medicine fall into
this trap. "The Lord's healing power works!" they proclaim; "Just
last year, our church cured twenty people of cancer by the power
of prayer alone!" Sounds impressive; but if the *base rate* of cancer
remission is 5%, and the church prayed for 400 patients, then
the statistical likelihood is that twenty patients would have gotten
better anyway. The results of the church's activities, that is, have no
effect on the actual rate of remission.

Bob's reasoning has gone askew, because he has not taken into
account the *base rate* of the disease. The chance that he's got the
disease is very low: only one in a thousand. Instead, Bob is relying
on the *specific* information at hand, namely that he's tested positive
for the disease; he thinks that the chances of being a false positive
are likewise only 5%. However, if he had taken into account that the
test's rate of failure is far higher than the rate of people who actually have the disease, he would have realized that a false positive is
much more likely than an accurate result.

We see the same mistake with the faith-healing example. The
base rate of cancer remission is 5%; the statistical likelihood is,
therefore, that twenty out of four hundred patients will get better
naturally. The church's rate of success doesn't improve the base
rate of remission, and therefore there is no evidence that prayer is
effective in treating cancer.

THE COMEBACK

The only comeback to this fallacy is to have a sound grasp of statistics and probability theory: you need to be able to point out not only that your opponent is committing this fallacy, but also specifically what's at stake here. However, during the 'fast and furious' process of debate it can be very difficult to spot these errors, let alone explain them to your audience.

SIGNIFICANCE

This fallacy illustrates some of the problems that we face when thinking about probability. It's worth thinking through the above example, to see how it works.

BEGGING THE QUESTION

Petitio Principii

Formal	Proponent A justifies *P* on the grounds that *Q*, and justifies *Q* on the grounds that *P*.

An argument whose premises assume the truth of its conclusion.

EXAMPLE

> *"People's motivations are always self-interested. Whenever they aim at another's welfare, they're really aiming at their own good. Hence, we see that people are always self-interested."*

REAL-LIFE EXAMPLE

In Molière's play *The Imaginary Invalid* we witness a group of physicians explain why opium puts people to sleep. They explain this by its *virtus dormitiva* (Latin for *'soporific capacity'*). But this just means that opium has the power to put people to sleep. So, the learned doctors effectively claim that opium puts people to sleep because it has the capacity to put people to sleep. But we already knew that opium has the power to put people to sleep–that's what we're trying to explain!

THE MISTAKE

An argument justifies its conclusion on the basis of its premises and its premises on the basis of its conclusion. Since the conclusion is precisely what's at stake, the premises must be acceptable without assuming the truth of the conclusion–otherwise, there's no more reason to accept the premises than the conclusion. But *question-begging* arguments violate this stricture: the truth of the premises depends on the truth of their conclusion. Such arguments are hopeless, as they assume what they set out to prove.

How do such arguments arise? The simplest way is when the conclusion simply rephrases the premises, as in the Molière example above. In these cases, the argument *appears* to be '*P*, therefore *Q*,' but is in fact '*P*, therefore *P*': '*P*' just rephrases '*Q*.'(This is particularly risky when the premise is rephrased into Latin. As we know, speaking Latin makes you appear profound; after all, "Quidquid latine dictum sit, altum videtur").

These arguments also arise when the truth of the premises depends *indirectly* on that of the conclusion. So, if I justify *P* by *Q*, but justify *Q* by *P*, I haven't justified *P* at all. In our 'altruism' example, the proponent justifies his belief that everyone is selfish by claiming that apparently altruistic acts are, in fact, indicative of disguised self-interest. But nobody would believe the second proposition without believing the first. Hence, the proponent of the argument has *begged the question*.

THE COMEBACK

This fallacy, when spotted, obviously makes the argument hopeless. The tricky bit is spotting it; after all, your opponent will, wittingly or unwittingly, be concealing his maneuver. Hence, you have to show that your opponent's premises assume the conclusion, either by rephrasing it or by presupposing it.

How you go about this depends, of course, on the argument. With the 'altruism' example, you might argue thus:

> "*People are always selfish!*"
> "*Well, I don't know: Lucy gave some money to a panhandler just now . . .*"

"Yeah, but she's obviously doing it out of selfish motives, like appeasing her conscience or something."

"Why would you think that?"

"Geez, didn't you hear me? Because people are always selfish!"

It's now obvious that your opponent has begged the question.

SIGNIFICANCE

This fallacy's name is a mistranslation of the Latin *petitio principii*, meaning to *assume the initial position* (in Greek: *to ex archês aiteisthai*). The term 'to beg the question' is sometimes used to mean '*raise*' or '*evade* the question,' which is, strictly speaking, inaccurate.

Formally, the argument is impeccable: it is of the form '*P*, therefore *P*,' which is logically impeccable, albeit uninformative. Particular types of this fallacy include *Circular Reasoning* (see p. 102) and *No True Scotsman* (see p. 145).

BIASED SAMPLE

Informal	Population M has a sub-class m, which has characteristic *P*. It is then inferred that M also has characteristic *P*. However, m is not representative of M.

Where a general conclusion about a population is drawn from the behavior of a small sample, when the sample does not accurately represent the population as a whole.

EXAMPLE

"A poll of 100 Americans showed that 98% believed in God. Hence, 98% of Americans believe in God." The pollster did not mention that the sample consisted of congregants of the First Pentecostal Church of Alabama.

REAL-LIFE EXAMPLE

A famous picture of President Harry S. Truman shows him brandishing a copy of the *Chicago Daily Tribune* with the banner headline "Dewey Defeats Truman," a few days after Truman had defeated Dewey in the 1948 election. How could the newspaper get it so wrong? Well, they went by their polls. But their polls were conducted by *telephone*, which, at the time, only rich Americans could afford. Hence, the polling sample was biased (albeit unintentionally).

THE MISTAKE

Sampling is a species of *inductive* reasoning: the sample is a micro-cosm of the whole population; what's true of the sample is taken to be true of the whole. In a *biased sample*, however, the sample doesn't represent the whole. So, in the first example, what's true of the members of the church was taken to be true of all Americans; however, this inference would only hold if *all* Americans were members of the First Pentecostal Church.

THE COMEBACK

You have to show that the sample does not represent the popula-tion. So, in the case of the first example, you could point out that the pollster's 'sample' only included Americans who lived in one particular location and subscribed to one particular worldview.

SIGNIFICANCE

This fallacy isn't always committed deliberately: it frequently comes about when sloppy polling methods are used.

Suppose an online poll says that 60% of Americans will vote for Hillary Clinton: does that mean that Clinton will cruise to victory in 2016? Not necessarily. After all, not all Americans have access to the Internet; of those who do, not everyone is interested in filling out online polls. Now, it could be the case that Clinton supporters all have Internet access and spend time participating in online polls, whereas Trump supporters generally don't. If that is the case, this particular sample does not represent America's voting intentions. So, Trump may gain a surprise victory, in spite of this poll's predictions.

BLIND AUTHORITY

Informal	Authority A states that *P*. Therefore, *P*.

Justifying an argument based on the say-so of an authority whose credentials have neither been examined nor questioned.

EXAMPLE

> *"My cult leader says that God will immolate the earth next week. Hence, the end of the world is nigh!"*

REAL-LIFE EXAMPLE

The beliefs and practices of cults provide clear, if distressing, examples of this phenomenon. For instance, the leader of the Heaven's Gate cult persuaded his followers that their souls could evacuate the earth and board an alien spacecraft supposedly trailing the Hale-Bopp comet, thereby reaching an above-human level of existence. The cult's members believed that death was a prerequisite for transcending their earthly existence, so 39 of them committed mass suicide in March 1997. Why did they do so? Answer: because the cult's leader told them to. The cult's adherents never raised the question as to *why* they should believe him.

THE MISTAKE

An *Appeal to Authority* (see p. 43) is legitimate only when the authority has a well-established expertise on the matter at hand. But the fallacy of *Blind Authority* ignores this criterion: the 'authority' is trusted blindly, without first establishing his credentials related to the matter at hand. Once we realize that, it becomes

apparent that there is no reason to trust this 'authority' more than anybody else.

THE COMEBACK

In an ideal world, it should suffice to point out that your opponent's appeal to authority is entirely ungrounded. However, this won't necessarily convince him; your appeal to reason may have no effect against his faith in his chosen authority.

SIGNIFICANCE

This argument is a fairly extreme type of the *Appeal to Authority* (see p. 43). But it's interesting to note how our unthinking obedience to authority makes us prey to this fallacy. *The Illuminatus! Trilogy* (a cult sci-fi/conspiracy novel) mentions the alleged anarchist pastime of the so-called 'Bavarian Fire Drill,' where anarchists would dress up in uniforms, stop the traffic, and, announcing a 'Bavarian Fire Drill' would make drivers exit and run in a circle around their cars. The drivers obey, simply because they are told to do so. (Of course, this is what the book tells you–should you believe this actually happened?)

CHERRY-PICKING

Informal	Evidence E supports *P*, evidence F contradicts it. Proponent A appeals to evidence E to prove that *P*, while ignoring evidence F.

Establishing a conclusion by means of evidence, but selectively citing only evidence that supports your conclusion, while suppressing any evidence which contradicts it.

EXAMPLE

"This study showed that a hundred cancer sufferers who used homeopathy recovered. So it is established that homeopathy can cure cancer!" (While failing to mention that the study had ten thousand subjects, the rest of whom did not respond to this 'cure').

REAL-LIFE EXAMPLE

The National Rifle Association frequently claims that the Constitution guarantees citizens the right to bear arms, citing the Second Amendment: "the right of the people to keep and bear Arms, shall not be infringed." However, they omit the first part of this clause: "A well regulated Militia, being necessary to the security of a free State," a phrase which casts the meaning of the Amendment in a rather different light.

THE MISTAKE

The proponent's argument looks legitimate enough: he makes a claim, and backs it up with some relevant evidence. However, he

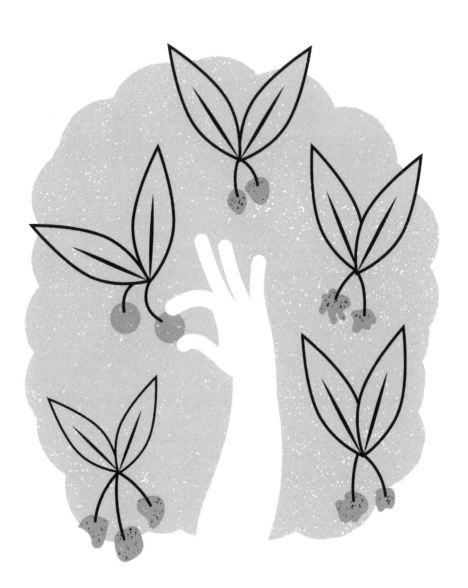

has failed to mention some *other* evidence, which contradicts his claim (as with the homeopathy example), or which qualifies his claim (as with the Second Amendment example). If *all* the evidence had been presented, his claim would be shown to be unjustified.

THE COMEBACK

The best comeback is to simply know more about the topic than your opponent. For instance, if your opponent is an NRA member quoting the Second Amendment, simply point out that your opponent is selectively quoting the text.

If you can't do this, you can still try and deconstruct the evidence adduced. So, with the homeopathy example, you could ask about the size of the study group, and the rate of natural remission from cancer. If it turns out that one in a hundred cancer patients will undergo natural remission, then his 'evidence' has proven nothing: we'd expect that a hundred patients out of ten thousand would spontaneously get better anyway.

SIGNIFICANCE

This fallacy is committed involuntarily as much as it is carried out voluntarily. We all subconsciously seek out evidence that corroborates our beliefs, and tend to ignore evidence that contradicts them–this is a phenomenon known as *confirmation bias*. So we'd be well advised to be on guard against sloppiness in our own thinking as much as our opponents'.

Sample Bias (see p. 93) can be an instance of this fallacy.

CIRCULAR REASONING

Informal	P is justified by Q. However, Q could only be justified by accepting P. (Alternatively: P is justified by Q, which is justified by a number of other steps, which are ultimately justified by accepting P).

Arguing for a conclusion on the basis of a set of premises, where the truth of the premises assumes the truth of the conclusion.

EXAMPLE

"My cult leader is infallible; he tells me so. I know what he tells me is true, because he's infallible."

REAL-LIFE EXAMPLE

Galileo Galilei's overthrow of the Aristotelian world-system was not, as is sometimes believed, achieved by simply dropping two different weights off the Tower of Pisa. Rather, he took apart Aristotle's arguments for his cosmological system one-by-one, showing that they were logically flawed. Galileo reconstructs Aristotle's argument for the earth's immobility thus:

P1. *Freely falling objects fall in a straight line towards the centre of the earth.*

P2. *The earth rotates, so the tower rotates around the centre of the earth.*

P3. *If the tower were rotating around the centre of the earth, the stone would not fall at the foot of the tower, but would fall to the west of it.*

P4. But the stone falls to the base of the tower.
∴ *The earth does not rotate.*

The problem with this argument lies in premise *P1*, which is precisely the point at issue. Sure, if you're an Aristotelian, you believe that the stone moves in a straight line towards the center of the earth. But if you're not, you can quite happily reject the view that the stone has a straight motion, and instead believe that it has a *compound* motion, comprised of its own motion and that of the earth's, which would account for it falling to the base of the tower. So, *P1* in the argument depends, ultimately, on the premise that the earth does not rotate: we can say that freely falling objects fall in a straight line only by assuming that the earth stands still. But that, of course, is precisely what the argument sets out to prove. So the argument (like the Earth's rotation) is circular.

THE MISTAKE

This fallacy is a subspecies of *Begging The Question* (see p. 90), and suffers from the same problem. Your opponent's argument must argue his contentious conclusion on the basis of uncontentious, or at least less contentious, premises. However, if the premises themselves are justified only on the basis of the conclusion, the argument can't work: there is no more reason to accept the premises than there is to accept the conclusion.

THE COMEBACK

It should suffice to point out the circuitous reasoning: since your opponent has given you no more reason to believe his premises than his conclusion, you've no reason to believe either. The difficulty, of course, lies in bringing the circularity to light, since it will usually be concealed. Of course, even if you point the circularity out, this may not help move the debate forward: I doubt that the cult member in the above example would be especially moved by your rational argument.

As mentioned, this is a subspecies of *Begging The Question* (see p. 90). However, it also differs from it, because here the premises and the conclusion must be different–the premises cannot be just a reworded version of the conclusion.

It's interesting to note, however, how much of our everyday life depends on this sort of reasoning. Economics gives us some good examples. Why is gold valuable? Answer: because people value it. But why do people value it? Answer: because it is valuable.

COMPLEX QUESTION

Many Questions or *Loaded Question Fallacy;*
Plurium Interrogationum

Informal	The speaker asks a question, which presupposes a number of facts P, Q, R, to which the respondent is not committed.

The speaker poses a question that contains a complex presupposition. *The presupposition is not stated, but is required for the question to make sense.*

EXAMPLE

Suppose someone asks me: "Have you stopped beating your wife?" The question requires a yes/no response; however, this requirement presupposes a number of facts, such as that I have a wife, and that I have, at some point, beaten her. If these aren't true, then I can't answer the question posed: I can't say yes, because I never beat my wife in the first place; I can't say no, because I'm not presently beating her.

REAL-LIFE EXAMPLE

The inimitable Stephen Colbert conducted a survey of peoples' appraisal of G.W. Bush's presidency, asking: "George W. Bush: *great* president, or *greatest* president?" This question, of course, presupposes that the respondent thinks positively of Bush, Jr.'s presidency in the first place.

THE MISTAKE

Any question demands a response, and will limit the sort of responses that are acceptable. So, a 'yes/no' question–did you, or did you not?–requires that the responder say either 'yes' or 'no.' A *Complex Question* is one where *both* of these responses presuppose something which the respondent does not accept. So, in the wife-beating example, both the 'yes' and the 'no' responses presuppose that the respondent has a wife, and has, at some time, beaten her; however, both of these presuppositions may be false.

THE COMEBACK

The trick here is to question the question. You have to show that the question is illegitimate, because it tries to sneak in presuppositions that you reject. So, if someone asks you "Have you stopped beating your wife?" you can turn around and ask: "Why do you assume that I have ever beaten my wife?" or even "Why do you suppose that I even *have* a wife?"

SIGNIFICANCE

This is a venerable fallacy, known in antiquity as the *Fallacy of the Horns*. The classic example (as put forward by a certain Eubulides the Eristic, a contemporary of Plato's) was the question "Have you lost your horns?" Answer 'no,' and you imply that you currently have horns; answer 'yes,' and you imply that you once did have horns, but have now lost them. Hence the phrase, "impaled on the horns of a dilemma."

EQUIVOCATION

Amphiboly

Informal	A term common to the premises and conclusion has two distinct meanings, such that the first meaning is required for the premises to be true, but the second meaning is needed for the conclusion to logically follow from the premises.

When the conclusion of an argument seems to follow from the premises, but only by virtue of an ambiguity in the meaning of the words used in the premises and conclusion.

EXAMPLE

Three and *two is five. Three is odd, two is even.*
Hence, five is both *odd* and *even.*

REAL-LIFE EXAMPLE

In the *Lord of the Rings*, Eówyn, doughty noblewoman of Rohann, is faced with the dreaded Witch-King, Lord of the Nazgul, who proclaims, "No man can kill me!" Unfazed, Eówyn, proclaiming "I am no man!" proceeds to slay him.

This exchange rides on the ambiguity of the term 'man,' as meaning 'member of the species *homo sapiens,*' on the one hand, or '*male* member of the species *homo sapiens,*' on the other. Nazgul's thought-process can be formalized thus:

P1. No man can kill me.

P2. Eówyn is not a man.

∴ Eówyn cannot kill me.

His conclusion follows from his premises. However, if we disambiguated the term 'man' as used in the premises, the argument would
run thus:

P1. No male *can kill me.*

P2. Eówyn is a human.

∴ Eówyn cannot kill me.

But spelled out thus, the conclusion patently *does not* follow from
the premises; had he realized this, the Witch-King might have
avoided being slain.

THE MISTAKE

A word is *equivocal* when it has two or more distinct and unrelated
meanings. For instance, the term 'bank' can mean 'the edge of a
river' or 'a place to deposit money' (if I went to the *bank* to deposit
a check, I wouldn't think to wear my galoshes lest I got my feet
wet). Fallacies of this sort trade on this verbal ambiguity: the term
means one thing in one of the premises, but something entirely
different in the other premise or the conclusion. But if the term is
ambiguous in this way, the conclusion can't logically follow from
the premises: the term on which the argument turns refers to different things in each of its uses.

THE COMEBACK

You should first point out the ambiguity of the term, and then
show how the argument works just because of this ambiguity:
the premises are only acceptable when the term is taken to mean
one thing, but the conclusion follows only when the ambiguous
term is taken at its other meaning. So, in the mathematical example above, 'and' has two meanings: in the premises, it denotes the

mathematical operation of *addition*; in the conclusion, it denotes the logical operation of *conjunction*. Thus disambiguated, the argument runs:

> P1. *Two* added to *three makes five.*
> P2. *Two is even, three is odd.*
> ∴ *Five is both odd and even.*

But that doesn't look like an argument at all. Which is good, because it isn't.

SIGNIFICANCE

An equivocation is a kind of ambiguity where the ambiguity is semantic, that is, where it trades on the word itself having two meanings. Other kinds of ambiguity are syntactic; that is, they stem from the fact that the word's meaning depends on the context in which it is used.

Verbal ambiguity is one of the great joys of natural language, and the wellspring of crossword puzzles, wordplay, and jokes ("My dog's got no nose!" "How does he *smell*?" "Awful!"). But this sort of ambiguity is best left to the music hall, not the debating chamber.

FAKE PRECISION

Formal or Informal	Argument *P* is supported by quantitative evidence E, where E lacks the quantitative precision needed to legitimately support *P*.

Supporting an argument with numerical data that appears to be more precise than it actually is.

EXAMPLE

> *"70% of the statistics you read in books are just made up by the author!"* [5]

REAL-LIFE EXAMPLE

A speech by Robert Kennedy (then Attorney General) in 1961:

> *Ninety percent of the major racketeers would be out of business by the end of this year if the ordinary citizen, the business man, the union official, and the public authority stood up to be counted and refused to be corrupted.*

We can't doubt the nobility of this statement. But *ninety* percent? Where did that figure come from? Had Kennedy conducted a poll of mob leaders? [6]

---

5 Withey, *Mastering Logical Fallacies*, p. 111

6 The example is found in Walton, *Informal Logic*.

Quantitative data is extremely useful in getting precise information about the case at hand. However, this quantitative data can only be so precise: even in engineering, measurements only admit of accuracy up to a certain degree. *Fake precision* occurs when this caveat is ignored–the data is presented as demonstrating something, without acknowledging that the data is only accurate within certain limits.

In some cases, the data simply can't be expressed in a precise quantitative way. In particularly bad cases (as with the two examples above), statistics are used to give a claim an aura of rigor, even though these supposed statistics are just plucked out of thin air.

THE COMEBACK

You need to challenge the methodology behind the adduced statistics. In the cases above, the response would be simple: the assertions that 70% of quoted statistics are invented, or that 90% of crime bosses could go out of business within a year if the people stood up to them, have absolutely no grounding in actual research. Alternatively, you could point out that the matter can't be expressed quantitatively at all; or else, that the measure your opponent uses just isn't that precise.

SIGNIFICANCE

This is a form of *Lying with Stats* (see p. 134). A lot of urban myths (*Appeal to Authority* [see p. 43] or *Appeal to Common Belief* [see p. 47]) have this flavor: "In New York, you're never more than six feet away from a rat!"; "Men think about sex every six seconds." Such pronouncements *sound* impressive, but are usually inaccurate: if men thought about sex that often, how on earth would they ever get anything done?

This fallacy has a particularly pernicious effect on policy and business, in cases where managers try to impose metrics and targets on things that can't precisely be measured. For example, in academia careers can stand or fall depending on the number of publications (or 'outputs,' to use the jargon) a scholar has to his credit, a metric which will favor a scholar who has put out ten mediocre papers to another who has put out a single brilliant one.

FALLACY OF COMPOSITION

Informal	Whole W is comprised of parts p1, p2, . . . , p*n*. Since each of the parts has a certain property, it is inferred that the whole has that property.

Inferring that what is true of the parts of a whole is also true of the whole.

EXAMPLE

> *"If I stand up during a football match, I can see better. Therefore, if* everybody *at the match stands up,* everybody *will see better."*

REAL-LIFE EXAMPLE

Homer Simpson is tasked with designing a car. He demands: "You know those colored balls they have on car aerials, so you can find your car in a parking lot? *Every* car should have one of those!" He forgets, of course, that these balls only help you find your car when a *few* cars have them: they would be of no help if *every* car were to have one.

THE MISTAKE

It is invalid to infer that the *whole* will have just those properties that its parts have: the parts will, when taken together, have properties that each individual part doesn't have. We can see this most simply when it comes to counting: Socrates and Hippias are each *one*; however, it does not follow that Socrates and Hippias are, *taken together*, one. Similarly, we can't infer that, just because each straw in a bale of hay is light, the entire bale of hay will be light: the bale of hay will be very heavy.

This fallacy is particularly tricky when it comes to *modal* statements (statements which deal with possibility). Suppose I say "There's no excuse for being unemployed: there's always somebody who needs their windows cleaned." Even if it is true that *any* unemployed person could get a job cleaning windows, it doesn't follow that *every* unemployed person could do so: there's only so many windows that need cleaning.

THE COMEBACK

You have to show that the properties of the whole can't simply be reduced to those of its parts. The examples above will help: we can't infer that, just because each straw is light, a whole bale of hay will also be light; add enough straw, and you end up breaking the camel's back. Similarly, you could point out that something being true of *any* member of a group doesn't imply that it's true of *every* member of the group: so, it's true that *any* person could have been the eldest sibling of their family; however, it's not true to suggest that *every* person could be the eldest sibling.

SIGNIFICANCE

This fallacy was again noted in antiquity; the Socrates/Hippias example is found, for example, in Plato's *Hippias Major*.

The converse of this fallacy is the *Fallacy Of Division* (see p. 117), where it is fallaciously inferred that the properties of the whole are also those of the parts: so, because the bale of hay is heavy, each blade of straw must be heavy. Both fallacies are kinds of category mistake.

FALLACY OF DIVISION

Informal	Whole W is comprised of parts p1, p2, and p3 . . . . Whole W has property P. Hence each of the parts will also have property P.

Assuming that what is true of the whole *is also true of each of its* parts.

EXAMPLE

"This bale of hay is heavy. The bale of hay is comprised of straw. Hence, each stalk of straw must also be heavy."

REAL-LIFE EXAMPLE

This fallacy is committed frequently in conversations about social justice. White males, as a demographic, are socially privileged; it is then inferred that *each* white male will have this 'privilege,' regardless of his personal circumstances. By this logic, the life of a white unemployed, homeless war veteran should be deemed *privileged* compared to that of a minority female CEO.

THE MISTAKE

This is the flipside of the *Fallacy Of Composition* (see p. 115), and shares the same flaw: what's true of the whole is not thereby true of *each* of its parts. A whole comprised of parts can have properties that none of these parts have; so, to use the Socrates/Hippias example, Socrates and Hippias are, *as a pair*, two; however, it clearly doesn't follow that Socrates and Hippias are *each* two.

You need to show that the whole's properties only belong to it precisely *because* it is a whole, and as such more than just the sum of its parts. The bale of hay example is simple enough: the bale of hay's weight is the *combined* weight of *all* the stalks of straw it comprises; but it doesn't have this weight because *each* of its parts has that weight.

FALSE ANALOGY

Informal	A is P, B is P. A is Q, therefore B is Q.

An analogy *is established between two things, A and B. A and B both have the characteristic P; A has the characteristic Q; hence it is inferred that B also has the characteristic Q.*

EXAMPLE

"An apple is a fruit, and it is round. A pear is a fruit. Therefore, a pear is round."

REAL-LIFE EXAMPLE

Electronic cigarettes allow smokers to enjoy the sensation and pleasurable effects of smoking without any of its detrimental effects to health and wellbeing. However, some authorities have tried to impose the same restrictions on their sale and use as with normal tobacco products, prohibiting, for example, their consumption in public places. They justify these strictures on the grounds that e-cigarettes are like normal cigarettes, ignoring the fact that e-cigarettes lack the harmful properties that normal cigarettes have.

THE MISTAKE

An *argument from analogy* tries to establish something about an *unknown* or *contested* case from something about a *known* or *uncontested* case. The rationale is that if you can establish that a certain thing holds of the uncontested case, you can infer that something similar holds of a similar but contested case.

Of course, we cannot readily assume that *anything* about the uncontested case holds of the contested case—if so, the two cases would be identical. However, an *extended* analogy ignores this stricture, maintaining that *anything* holding of the uncontested case must also hold of the contested case.

THE COMEBACK

Since your opponent's argument rests on the apparent *similarity* of the cases, you need to show that the cases are in fact *dissimilar*.

So, with the smoking case, your opponent argues thus:

P1. *Cigarettes should be restricted.*
P2. *E-cigarettes are similar to cigarettes.*
∴ *E-cigarettes should be restricted.*

The fault here lies in step P2. E-cigarettes may be similar to normal cigarettes, in that they deliver nicotine via inhalation and look sort of similar to normal cigarettes. However, they are also *dis*similar, in that they are not harmful (or, at least, there is no evidence yet for their being so). But cigarettes are restricted precisely *because* they are harmful. Hence, the similarity alluded to in premise P2 is not extensive and substantial enough to justify the conclusion.

SIGNIFICANCE

Analogies are extremely useful, not only in politics, but in science and philosophy as well: by reflecting on an uncontentious case, we can more easily decide something about a similar, contested case. The contentious point, however, is establishing *how* the two are similar. If A is P and Q, and B is P, it does not yet follow that B is Q. For that, we would have to establish that A is Q *because* it is P; if so, since B is P, it must also be Q.

Some *Slippery Slope* (see p. 175) arguments may be fruitfully analyzed in these terms: because A is permissible, and A_1 is similar to A, it follows that A_1 is permissible; since A_2 is similar to A_1, and A_1 is permissible, A_2 is permissible . . . and so on, all the way to the final, potentially awful consequences of this kind of lax logic.

FALSE DILEMMA

Informal	Proponent A offers a choice between P or Q, on the condition that one, and only one, of the two must be chosen; in reality, however, accepting *both* P and Q, or a third alternative R, are also viable options.

A choice is presented between two alternatives. The proponent presents this choice as exhaustive *and* exclusive: *one of the options* must *be chosen; no third option is permitted or even entertained. However, in reality, these two options are neither exclusive nor exhaustive.*

EXAMPLE

"Either you're married, or you're a bachelor. You're not married; so you must be a bachelor! How about a drink?"

REAL-LIFE EXAMPLE

A recent newspaper headline declared that 90% of Icelanders were atheists. This finding was based on an online poll which asked the question: "Which do you believe created the universe: God or the Big Bang?" where readers could choose only one option.

THE MISTAKE

A *dilemma* occurs when two *exclusive* and *exhaustive* answers are presented as the only possible answers to a problem. The dilemma is *false* when the options are neither exclusive nor exhaustive—we may also conceivably take both options, or take an altogether different third option. So, in the bachelor example, it doesn't follow that a man is a bachelor *just because* he's unmarried: he could be a widower, in a civil partnership, a monk, a child, and so on. The question presents its two options as *exhaustive* when there are in fact other options.

The case of the newspaper poll is even more egregious. First, it is plausible, if unlikely, that someone may think that the universe was created by something else altogether; for instance, the respondent may be a Gnostic, and believe that the world was created by a demon; alternatively, he may not believe the universe *had* a beginning.

However, the more obvious fault is that the poll presents the two options as being *exclusive*: *either* you believe God created the world, *or* the Big Bang; if you think God created the world, you don't believe the Big Bang did. But any intelligent Christian would conceivably believe in *both*; he would reconcile science with his faith by believing that God created the world *via* the Big Bang.

THE COMEBACK

Show that the dilemma is artificial–that you can choose a third option, or you can take both options together. So, in the first case, you could respond: "I'm not married, but I'm not a bachelor either: I'm a monk!" (thereby denying that the presented choices are *exhaustive*). In the second case, you could say: "I believe the Big Bang created the world, but that *it* was caused by God" (thus denying that the presented choices are *exclusive*).

Your opponent's linguistic usage may complicate things. Thus, in the bachelor case, your opponent's argument may seem plausible: if 'bachelor' just means 'unmarried man,' and a man must be either married or unmarried, you may think that a man must be either *married* or *a bachelor*. But not all unmarried men can really be described as bachelors: a child is unmarried, but it's hardly apposite to say he's a bachelor.

The *False Dilemma Fallacy* (see p. 122) frequently gives rise, in turn, to the *Complex Question* (see p. 106) fallacy. So, the question "have you stopped beating your wife?" allows two responses: *yes* (I used to beat my wife, but now I've stopped), or *no* (I still beat her). It ignores the possibility that I'm unmarried, or that, albeit married, I've never laid a finger on my wife.

To avoid this dilemma, it's important to grasp the distinction between *contradictory* and *contrary* properties. Two properties are *contrary* when something cannot have *both* properties; they are *contradictory* when anything must have either one or the other. So, 'black' and 'not-black' are *contradictory* properties: if something isn't not-black, it's black. 'Black' and 'red,' however, are *contrary* properties: nothing can be both black and red, but it doesn't follow that, if something's not black, it's red (or indeed any other particular color). The false dilemma occurs when contrary properties are presented as contradictories.

HASTY GENERALIZATION

| Informal | Each instance of a small sample of thing A has the property X. Hence, all instances of A have property X. |

A general rule about something is inferred from a few instances of that thing.

EXAMPLE

> *"I drink a bottle of whiskey a day, but so what?*
> *My grandpa drank that much, and he lived to be*
> *ninety-seven, and danced a jig every day!"*

REAL-LIFE EXAMPLE

Bertrand Russell gives the example of an intelligent chicken who is fed every day by a farmer. The chicken thinks, "every time the farmer visits, I get food!" Unfortunately, one day, the chicken doesn't get fed, but instead gets his head on the chopping block. The chicken made a generalization about the farmer's behavior; sadly, this generalization was too hasty.

THE MISTAKE

A *hasty generalization* is an example of *inductive reasoning*: you observe a few cases that have a property in common, and infer that every similar case will share that same property. The more cases at your disposal, the more secure the inference. A *hasty generalization* is an inductive inference based on too few instances: the sample size is too small to support a general conclusion. The presence of the property in question in the small sample may simply be a statistical anomaly, or the presence of the common trait may be explained by another variable.

THE COMEBACK

You could argue that the sample size your opponent has taken is too small to warrant his general conclusion. However, you would be on more secure ground if you could marshal some statistical evidence to show that your opponent's conclusion doesn't hold as a general rule. So, with the whiskey case, statistics show that, generally speaking, heavy drinkers die much earlier than the average person; your opponent's grandpa may just have been one of the lucky alcoholics who survive to old age.

Your ground would be even more secure if you could show that your opponent's inference didn't take certain variables into account. True, his grandpa lived to a ripe old age despite his drinking; but he had a certain gene X, which makes the liver resilient, but is present in only 1% of the population.

SIGNIFICANCE

Generalizations do have their place: if I'm in the savannah, and see two of my friends being gobbled up by lions, I should probably conclude that lions are to be avoided. But not always: if I'm in New York, and one of its denizens is rude to me, I shouldn't conclude that *all* Americans are rude. Generalizations should utilize a large sample set, ideally with controlled variables; although, as the lion example shows, this can only be a general rule.

JUST BECAUSE

Mother Knows Best

Informal	Proponent A justifies proposition/command *P* solely on the grounds that it is *his* assertion.

Proponent A justifies command or assertion P by simply positioning himself as an unquestionable authority on P.

EXAMPLE

"Don't eat that jam!"
"Why?"
"Because I said so!"

REAL-LIFE EXAMPLE

The Declaration of Independence reads thus:

We hold these truths to be self-evident, that all men are created equal, that they are endowed by their Creator with certain unalienable Rights, that among these are Life, Liberty and the pursuit of Happiness . . .

Stirring words. But what is the *argument* by which they establish this substantive conclusion? The text itself refuses to provide any, establishing it by fiat. (Admittedly, however, going into a disquisition about Locke and Montesquieu would rather spoil the effect of the sentence!)

THE MISTAKE

Arguments or commands require justification: the person uttering either must provide a reason why we should do or accept what he says. But this move doesn't just fail to provide a justification, it repudiates this very requirement–the proponent refuses to justify his assertion at all, claiming he has a right to make assertions by fiat. Needless to say, this appeal hardly commends itself to his opponent's reasoning.

THE COMEBACK

The response is simple enough: you should politely, but firmly, demand a better justification than your opponent's say-so. Don't let your opponent bully you into silence.

SIGNIFICANCE

This is a form of *Appeal to Authority* (see p. 43); however, it makes no appeal to the authority's expertise. The proposition isn't justified because the alleged expertise of an authority makes it more likely that it's true; instead, it is considered justified simply because someone commands assent to it.

Although the *Just Because* fallacy has no place in meaningful debate, it must be taken seriously in everyday life. This is especially so with, for example, someone's sovereignty over their body or property. So, if I say, "don't touch me!" this appeal does not in fact *need* any more justification than "because I said so." My authority in this instance doesn't rest on *expertise*; I'm simply asserting my rights over my own person.

Other forms of authority can have this absolute character: a soldier, for example, is expected to obey his superiors unquestioningly. Again, in such contexts authority can command assent without being associated with expertise.

Other decisions can also legitimately rest on a 'just because.' In Britain, people drive on the left. But why should the law command one to drive on the left, rather than the right side of the road? (both are equally good directions). Answer: *just because*. You need to pick one side to drive on; both sides are equally good; the authority's decision is no less binding because it is arbitrary.

This is also a question of profound theological import. God commands the good; but does He command it because it is good, or is it good *because He commands it*? St. Thomas Aquinas takes the former view; however, the 'voluntarists' (Occam, Duns Scotus, and their followers) take the latter. On their view, morality rests on a *'just because:'* what is good is ultimately arbitrary, the result of God's contingent will.

LUDIC FALLACY

Informal	A model M is used to make predictions about a certain domain D. However, M is defined with strict parameters that are not always present in D.

Taking a model of reality to represent reality, forgetting that the model is predicated on parameters with which reality quite freely dispenses.

EXAMPLE

Nassim Taleb's Black Swan *gives this example: Two men—Bob and Bill—are walking through a dodgy area of town. Bob is nervous; Bill says: "don't worry, I have a black belt in karate!" Sure enough, a mugger approaches them, and Bill tries to fell him with a karate chop, as he has done many times in the dojo. The mugger just shoots him dead. In the strictly regulated martial competitions of the dojo, Bill's karate skills are formidable; on the street, not so much.*

REAL-LIFE EXAMPLE

Casinos make their money by precisely controlling the net profits from their games. If you gamble, sometimes you win, other times you lose; the house, however, always wins at the end of the day. This isn't because the casino is *lucky*, but because they have carefully controlled the various outcomes of their games: slot machines will pay out every so often; the roulette wheel usually pays out to the house; the blackjack dealer has an intrinsic advantage. They try to minimize the impact of patrons who play these odds: if you're caught counting cards, you'll be banned; if you're caught cheating, you might get your hands broken.

In 2003, The Mirage casino sustained a $100 million loss. Why so? Had a canny gambler played the odds and won? No: it was because Roy Horn, of the Siegfried and Roy duo, had been mauled by his white Bengal tiger during their show. Although the duo had performed with the tiger for several years, the animal apparently went mad and attacked him. The casino made a predictable profit from their games, but sustained a loss because of a random event.

THE MISTAKE

Models are a useful tool when making predictions: by simplifying the picture of reality, they make things easier to predict and explain. For instance, a physicist's model of Newtonian mechanics will assume a frictionless vacuum; even in a context where friction and air exist, it's just easier to model reality dispensing with these variables. Similarly, a casino can precisely model the profits it will make from its gamblers, and use these models to project its earnings.

However, reality will always comprise more variables than a model could possibly capture. So, the casino's models didn't factor in the possibility that Roy's tiger might maul him. This isn't just a matter of building a different, better model: *no* model could accurately predict each and every random incident like that.

The story of Bob and Bill has a similar moral. Bill is an effective combatant within the strictly controlled parameters of the dojo. When faced with an opponent who dispenses with these parameters, he's in trouble.

THE COMEBACK

If you suspect that this fallacy may be in play, chances are that you are debating something at a high level of sophistication. Your opponent is using a model to predict and explain certain facts about reality; your job is to remind him that reality isn't always as neat as his model–it will always comprise variables which his model ignores. Your opponent may, for instance, appeal to an economic model, which assumes that people are rational economic agents, always pursuing the course of action that they think will maximize their profit. Hint: they aren't.

SIGNIFICANCE

The name *Ludic Fallacy* comes from the word *ludus* (Latin for 'game'). Games are defined by their clear, unambiguous rules: certain moves are allowed, others are prohibited; there is a clear mechanism to determine the winner; and so on. Even in a game of chance, the odds are visible and well-defined: in a game of craps, I've got a one-in-thirty-six chance of rolling snake eyes.

But real life isn't a game–there aren't any rules that prohibit certain outcomes from coming to pass. You cannot, therefore, even hope for the precise, controlled randomness of a game of chance: you might beat your opponent in a hand of cards, but still lose your stake if he punches you in the face and runs off with the money.

LYING WITH STATS

Formal or Informal	Proponent A attempts to support argument P with statistical data S, where S does not support P.

Supporting your argument by using statistical data in a misleading manner.

EXAMPLE

> *"100% of male violence is committed by males.
> Therefore, 100% of males are violent."*[7]

REAL-LIFE EXAMPLE

The *Spectator*, a long-running British periodical, recently ran an article celebrating the government's educational policy, claiming that standards in state schools were on the rise. It used, as evidence for this, the claim that the 80 best state schools were better than the 80 best private schools.

They forgot to mention, however, that Britain has *many more* state schools than private schools: 500 private schools, as against 3,200 state schools. The relevant metric should have measured a similar *proportion* of each kind of school together: so, the top fifty private schools against the top three hundred and twenty state schools. Because they ignored this dimension, the top eighty state schools may compare favorably against their private counterparts just because they are a statistical outlier.

---

7 from *Viz* magazine

There is no one single mistake that people make when *Lying With Stats*. Sometimes, the error is just a simple arithmetical error: they can't do the math. At other times, as with the school example, the relevant metrics are misleading; for instance, comparing quantities that are *numerically* equal when they should be comparing quantities of similar *proportion*. Other times, they use a biased sample, or they use statistics to disguise completely bogus arguments (as with the male violence example). *Lying with Stats* can be coupled with lots of other logical fallacies; thus, the male violence example uses statistics to deftly conceal its *Affirmation of the Consequent* (see p. 29).

THE COMEBACK

Since there's no single fallacy here, there's no single comeback. You need to have a firm grasp of statistics, not only to be able to detect your opponent's error, but also to avoid being taken in by his apparent rigor. Your opponent isn't trying to back up his arguments with florid words and empty rhetoric; he is presenting as rigorous, he has *numbers* on his side. However, it turns out that numbers can be manipulated as easily as words.

SIGNIFICANCE

This fallacy encompasses a number of others, such as the *Biased Sample* (see p. 93) or the *Base Rate* fallacy (see p. 87). It can be combined with any number of others; indeed, it helps conceal them, giving bogus arguments an air of precision, and scaring away any unwanted inquiry. If you have some time to spare, you may want to pick a few fallacies from this book and try to make them sound more plausible by inserting some junk stats.

MAGICAL THINKING

Informal	Two events, E and F, are thought to be causally connected in a supernatural way.

Thinking that two events are causally related not because of any reason or evidence, but because of a presumed supernatural connection.

EXAMPLE

Any superstitious practice falls under the rubric of this fallacy. I trod on the pavement-cracks; later, my mother sustains a spinal injury; clearly, stepping on the crack brought bad luck.

REAL-LIFE EXAMPLE

If you've ever found a four-leaf clover and thought that it would bring you good luck, you've fallen for this logical fallacy.

A story about the physicist Niels Bohr provides an amusing twist on the fallacy. He is said to have had a horseshoe above his door, something traditionally thought to bring good luck. A guest asked him: "Surely you don't believe in this superstitious nonsense, Professor Bohr?" "Of course not!" Bohr responded, "but I'm told it brings good luck, *whether you believe in it or not.*"

THE MISTAKE

The world isn't run by magic. Four-leaf clovers don't bring you good luck, treading on cracks won't injure your parents, sacrificing a bull won't land you that job. Sorry, that's just not how the world works.

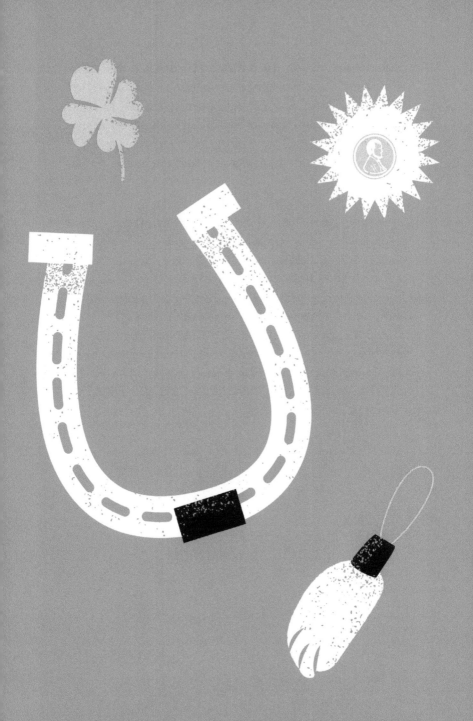

THE COMEBACK

You don't have to work particularly hard to respond to this: all you need to do is point out that there isn't any causal connection between the events your opponent is attempting to link. There isn't any scientific evidence that four-leaf clovers will ward off bad luck, or any statistically significant correlation between stepping on the cracks in the pavement and parental spinal injuries.

SIGNIFICANCE

So why is this kind of thinking so popular? We might think of it as a sort of *post hoc ergo propter hoc*: E happens after F, so F must have caused E. So (as the old rhyme goes), I step on a crack, then my mother breaks her back; naturally, I may infer, my stepping on a crack *caused* my mother to break her back. But we can't establish that F *caused* E just because E chronologically followed F.

Another explanation may be sought in the phenomenon of *confirmation bias*: people selectively look for evidence that supports their pre-existing views. A man finds a four-leaf clover; soon after, he gets the job, but loses his wallet. He attributes the success, but not the loss, to the clover.

MORALISTIC FALLACY

Informal	*P* ought to be the case. Therefore, *P*.

Thinking that something is the case just because it ought to be the case.

EXAMPLE

> *"All people should be equal. Therefore, one person cannot have any innately superior talents compared to others."*

REAL-LIFE EXAMPLE

The advent of Darwinism was an intellectual earthquake, not only in the field of biology, but in morality and theology as well: how can we assert that man is unique, if he is no more than a glorified ape? This moralistic reaction to Darwin's theory is memorably expressed in Benjamin Disraeli's speech in Oxford:

> *What is the question now placed before society with the glib assurance that to me is most astonishing? That question is this–Is man an ape or an angel? I, my lord, I am on the side of the angels.*

In other words, we should reject Darwinism, not because of empirical evidence, but because it doesn't fit with our moral sensibilities.

Another example can be found in the history of early Pythagoreanism (6th-5th centuries BC). The Pythagoreans believed that the world was governed by principles of number; indeed, that it was *made* of numbers (whatever that means!). They were extremely perturbed, therefore, when Hippasus, a member of their school, discovered that the square root of two was *irrational*, namely that it could not be expressed as a ratio between two natural numbers. Their cosmology thus besmirched, they condemned the unfortunate mathematician to death.[8]

THE MISTAKE

The world has no obligation to conform to our moral sensibilities, however high-minded they may be. Just because something *ought* to be the case, it doesn't mean that it *is* the case, or even that it *can* be the case—nature doesn't obey our moral sensibilities, absent our intervention; even then, it puts a hard limit as to what *can* be imposed on it. To think otherwise is a misplaced idealism, as absurd as the Pythagorean condemnation of Hippasus.

THE COMEBACK

The only secure comeback here is to marshal the empirical data against your opponent. But be warned, there is little in the field of sociobiology and evolutionary psychology that is not contentious, precisely because its critics fear its sociopolitical implications. In this context you should be especially careful about making an *Appeal to Authority* (see p. 43).

It may be better, then, to assure your opponent that, even if nature has these limitations, this doesn't mean that we can't improve our lot. Assume, for the sake of argument, that a study tells us that men are, on average, better mathematicians than women. This clearly doesn't imply that any *individual* woman will be mathematically inept: drawing such a conclusion would amount to committing the *Fallacy of Division* (see p. 117), and is quite easily refuted by considering the illustrious contributions of female

---

8   This story is reported in Diogenes Laertius, but as with anything to do with Pythagoras and his school, its truth must be taken with a grain of salt.

mathematicians. Still less does it mean that we should, for instance, restrict the study of mathematics to men.

SIGNIFICANCE

This is the inverse of the *Appeal to Nature* (see p. 67), or, as it is sometimes erroneously dubbed, the *Naturalistic Fallacy* (see p. 150). In the *Appeal To Nature*, the proponent tries to derive an 'ought' from an 'is': what holds in nature is what *ought* to hold. The *Moralistic Fallacy* is the converse of this, trying to derive an 'is' from an 'ought': the proponent assumes what *is* the case from what *ought* to be the case. Both fallacies need to pay better attention to the distinction between fact and value.

It has been noted that both left-wing and right-wing thinkers tend to fall prey to these fallacies. The right typically commits the *Naturalistic Fallacy*; so, from a study indicating that men tend to be better mathematicians, they are likely to conclude that women ought to stay at home and leave all the fancy math stuff to the men. The left, on the other hand, will typically commit the *Moralistic Fallacy* (throwing in an *Appeal to Anger* [see p. 40] for good measure): this study is *sexist*, so it must be flawed! Both sides would benefit from thinking about the fallacious nature of their reasoning.

MOVING THE GOALPOSTS

Shifting Sands

Informal	Proponent A puts forward argument P. Opponent B insists that evidence E is necessary for P to be accepted. Proponent A produces evidence E. Opponent B now demands a new, more stringent standard of evidence E_1 for P to be accepted.
	――――― OR ―――――
	Proponent A accepts E as the standard of proof for P, but relaxes the criterion of proof to evidence E_2 after realizing that the standard E_1 cannot be met.

To raise, or lower, the standard of proof required for accepting an argument, after the argument has been shown to meet, or fail to meet, a previously agreed-upon standard of proof. More generally, to change the terms of the debate or argument after the debate or argument has begun.

EXAMPLE

If we are to believe the movies, extortion relies on this sort of argument. "You've paid a million dollars for your wife, as we agreed; but if you want to see her again, you'd better pay a million more!"

REAL-LIFE EXAMPLE

Lawrence Krauss recently wrote a book stating that science had refuted the First Cause argument for God's existence. Contrary to the theologians' belief that "nothing comes from nothing," the famed cosmologist stated that science had shown that something–i.e., the universe–could indeed come from *nothing*. Critics pointed out that the 'nothing' from which the universe arose wasn't exactly *nothing*: it was a quantum vacuum, which, although containing no *particles*, is not *nothing*. To this, Krauss responded: "I don't really give a damn about what 'nothing' means to philosophers; I care about the 'nothing' of reality. And if the 'nothing' of reality is full of stuff, then I'll go with that." Here, it is the proponent of the argument who misses, and then proceeds to move the goalposts.

THE MISTAKE

This isn't necessarily a *fallacy*, but it *is* cheating. In any game, you need clearly defined standards to measure who wins and who loses; all parties should agree to these standards in advance. When an opponent *moves the goalposts*, he's cheating in just this manner: you've both agreed that you need to meet a certain pre-established standard of evidence; however, when you meet this standard, he changes the standard.

THE COMEBACK

The only effective comeback here is to complain that your opponent is cheating: you've both agreed on a standard of proof; by subsequently demanding a different standard of proof, whether more or less stringent, your opponent has violated this agreement. Feel free to call your opponent a jerk, an SOB, or a villain, depending on your mood.

This isn't really a *fallacy* as much as a gambit; it is to be eschewed not because it is invalid, but because it is unsportsmanlike. It gets its name from soccer, referring to the practice of widening or narrowing the goalposts to your team's advantage or your opponents' team disadvantage.

This move *can* encompass a number of *bona fide* fallacies, such as the *No True Scotsman* fallacy. Here, the proponent making a claim shifts the criteria of evidence for this claim by defining the term in question in such a manner that it cannot fail to satisfy his claim. For instance, the proponent claims that "no Scotsman takes sugar on his porridge." His opponent replies: "But my uncle Hamish is Scottish, and takes sugar on his porridge." The proponent retorts: "Ah, but no *true* Scotsman takes sugar on his porridge." By making his definitions more and more stringent this kind of opponent tries to render his argument immune from criticism.

MULTIPLE COMPARISONS FALLACY

Informal (Statistical)	Tests $\{T_1, T_2, \ldots, T_n\}$ are conducted to test hypothesis H. One test, T_m, shows some evidence that H is correct. Therefore, the results of the test are taken to confirm H.

Drawing significant statistical inferences from any *positive or negative results gleaned from tests conducted on a multiplicity of groups or criteria.*

EXAMPLE

Fat Tony, a hustler, has bought a hundred coins from Jimmy the Fence, and wants to know if any of them are crooked. In order to check them he flips each of them twenty times. Most are fine; but one coin flips heads nineteen times out of twenty. He concludes that Jimmy the Fence has sold him a crooked coin.

REAL-LIFE EXAMPLE

The rise of cellphones in the '90s was accompanied by fears that cell phone towers represented a health hazard. Swedish researchers investigated whether living near cell phone towers represented a health hazard, and found that people living near them had an increased incidence of brain tumors. Cue the headlines: "Cell Phone Towers Cause Brain Tumors!"

Samples can help us make wider inferences, but only within a certain margin of error. On average, a sample of population should, if selected well, represent the whole. However, this doesn't mean that *any* particular sample will do so—it's entirely possible that the selected sample is a fluke. Given a large enough sample base, this isn't likely, but remains a possibility. Hence, given a hypothesis, there is a certain probability that a sample will confirm that hypothesis not because the hypothesis is true, but because the sample is abnormal.

The mistake lies in using this freak sample to say something significant about the population as a whole; for example, to confirm or deny a hypothesis. So, flip enough coins, and it's likely that one will give a run of heads: Fat Tony shouldn't whack Jimmy the Fence just yet. Similarly, it's quite possible that a population living near a power line may suffer a greater-than-average rate of *some* disease, just by statistical fluke; look for enough diseases, and you can confirm anything is unhealthy.

THE COMEBACK

This fallacy is related to the *Base Rate* fallacy (see p. 87), and invites a similar response. You should remind your opponent that statistical sampling and hypothesis testing delivers accurate results only within certain parameters; given a large enough sample set, you will have samples that fall outside of the norm. Thus, given any hypothesis, there is a *base rate* at which a sample will confirm the hypothesis because of natural luck. You need to remind your opponent of these basic statistical facts, and encourage him to do more sampling.

Related to this fallacy is the *Texas Sharpshooter Fallacy*, describing a situation where a Texan unloads a gun onto a side of a barn, paints a bull's-eye over each of the bullet holes, and declares himself a sharpshooter. Given a large enough sample size, in other words, you can find all sorts of weird patterns; the fallacy arises when it is forgotten that these weird patterns are to be expected from this kind of sample size.

NATURALISTIC FALLACY

Informal	Identifying a natural property P with the good. Colloquially expressed: "P is 'natural', therefore P ought to be done."

Strictly speaking, this fallacy has to do with identifying a non-natural property, such as goodness, with a natural property, such as pleasure. More colloquially, deriving the fact that something ought to be the case from the fact that it is the case. More colloquially still: using standards derived from nature to determine what ought to be the case in human societies.

EXAMPLE

"Women are naturally suited to be homemakers. Therefore, all women ought to be homemakers!"

REAL-LIFE EXAMPLE

Many political opinions, both reactionary and revolutionary, depend on this fallacy. You may be familiar with it in, for instance, a homophobe's declaration that homosexuality is wrong, because it is *unnatural*; or the Fascist belief that in society, as in nature, the strong should rule and crush the weak. But it's also found in revolutionary rhetoric, as with Rousseau's adage "man is born free, but is everywhere in chains."

This argument endows 'Nature' with an unwarranted normative significance. This significance is unwarranted because, first, *nature* isn't always benign (get stung by a wasp and tell me otherwise); secondly, it's unclear that we can really distinguish what's 'natural' from what's 'unnatural' in this way, especially when it comes to man. After all, man is *naturally* able to talk, to invent things, to be social. So, how can any of his activities be *unnatural*?

THE COMEBACK

Several responses are possible (as with the *Appeal to Nature* [see p. 67]). First, you might argue that the behavior your opponent advocates is, in fact, entirely *natural*. So, it's hard to argue that homosexuality is unnatural, if animals also engage in it. Secondly, you could point out that what's natural isn't *ipso facto* good. So, even if women are, 'by nature,' better equipped to be homemakers (whatever that means), it doesn't follow that they ought to be restricted to the home, or denied careers. After all, we humans are pretty good at overcoming our 'natural' limitations.

It is worth noting Joseph de Maistre's response to Rousseau: "It would be equally correct to say that sheep are born carnivorous and everywhere eat grass."

SIGNIFICANCE

This fallacy's name was coined in G.E. Moore's *Principia Ethica*, where he takes issue with statements that identify goodness with a *natural* property such as pleasure. Moore thought these statements were problematic; although one can say that *pleasure is a good thing*, one cannot *identify* the property 'good' with something like 'pleasure,' since they are metaphysically different sorts of properties. This is the *Naturalistic Fallacy* in its proper sense.

The only debates, however, in which the fallacy has this meaning are arcane philosophical discussions–you will seldom, if ever, encounter this definition in the context of a political debate. In ordinary parlance, the term is usually a synonym for the *Appeal to Nature* (see p. 67), and is the converse of the *Moralistic Fallacy* (see p. 139).

NIRVANA FALLACY

| Informal | Proponent A puts forward a proposal P to solve a certain problem. Opponent B points out that P would not completely solve the initial problem, or would fail to solve other, related problems. Opponent B therefore rejects proposal P outright. |

Criticizing a proponent's solution to a problem on the grounds that it does not solve the problem completely; *in other words, on the grounds that it falls short of an* ideal *solution to the problem.*

> *"You say that banning drunk driving will save lives on the road. But people will still die in car crashes, whether driving under the influence or not!"*

REAL-LIFE EXAMPLE

Advocates of 'abstinence only' sex education are opposed to teaching teenagers about contraception on the grounds that condoms and other birth-control methods can, and do, fail. Since contraceptives don't *always* work, they contend, they cannot provide the solution to teen pregnancy. They ignore the fact that condoms and other contraceptives still have a very high success rate, and are therefore an effective tool in fighting teen pregnancy and STDs.

THE MISTAKE

You can make things better without making them perfect. (Indeed, if you aim at perfection, you will frequently fail to do anything good at all, as all perfectionists know!) Your opponent criticizes you for not making a situation perfect; however, your aim was not to make things perfect, but only to improve things as much as possible.

THE COMEBACK

You need only point out that your aim was never to make things perfect, only to improve things a bit (unless you did promise perfection, in which case you asked for it!). Your proposal won't bring about the Second Coming, but it may still ameliorate things a little bit–and isn't that better than nothing? You could also ask your opponent if *he* has an alternative solution to the problem; if he doesn't, then your solution may have limited effectiveness, but is still the best solution on the market.

SIGNIFICANCE

There are certain circumstances under which this sort of argument *can* be legitimate: where the proposal fails a cost/benefit ratio, for instance, or where it isn't as effective as a competing proposal. So, if the proposal only has limited effectiveness, it may not justify the cost of its implementation. Alternatively, if the opponent's proposal not only solves the problems that yours does, but also solves some problems that yours doesn't, then his solution should be preferred.

This fallacy is, in some ways, the flipside of the *Appeal to Desperation* (see p. 50). In that instance, a course of action was recommended because it minimally satisfied a certain demand, without considering the cost-benefit ratio, the feasibility of other plans, etc. Whereas that fallacy placed the bar for accepting a course of action too low, here the bar is set too high: unless the proposal can bring about heaven on earth, it should be rejected. Needless to say, both positions are immoderate.

NON SEQUITUR

Informal	An argument that states that *P, therefore Q,* when *P* does not in fact imply *Q*.

When one statement is presented as following from another, while it logically does not.

EXAMPLE

"If it's sunny, I garden. It's sunny. So, the diagonal of a square is incommensurate with its sides!"

REAL-LIFE EXAMPLE

An unconfirmed, but still illuminating, anecdote relates that Euler and Diderot–the former, Europe's greatest mathematician; the latter, the scourge of its political-religious establishment–were once engaged in a public debate about the existence of God. Euler supposedly opened with the gambit: "Sir, $(a+b^n)/n = x$; hence God exists, answer please!" To this, Diderot supposedly had no answer.

THE MISTAKE

The term *non sequitur* encompasses pretty much every other fallacy there is. The argument asserts that a conclusion follows from the premise, whereas it doesn't; each species of *fallacy* simply determines *in what precise manner* the presented conclusion is unwarranted. So, *Affirming the Consequent* (see p. 29) is technically a *non sequitur*, because the form of the argument indicates that the conclusion cannot follow from the premises; an *Appeal to Anger* (see p. 40) is a *non sequitur*, because nothing follows from the fact that I'm angry about something; and so on and so forth.

If the *non sequitur* is a fallacy in its own right, it will simply be one where the conclusion does not follow from the premise, and where this cannot be subsumed under another subtype of fallacy. The apocryphal Diderot-Euler exchange is an example of this: the theorem $(a+b^n)/n = x$ implies *nothing* about God's existence or non-existence. Euler didn't commit any specific kind of fallacy–the premise of his argument was simply irrelevant to its conclusion.

THE COMEBACK

You need to point out that the conclusion and the premises have nothing to do with each other. Typically, you will need to do so by recognizing what kind of fallacy you're dealing with, and responding appropriately. However, if the *non sequitur* can't be reduced to another fallacy (as with the Euler/Diderot example above), then you should ask your opponent to spell out his reasoning a bit more: exactly *how* does this theorem show that God exists?

The Euler/Diderot case does highlight a particularly pernicious use of this fallacy: cases in which your opponent's argument tries to blind you with 'science,' or, more accurately, by applying pseudo-scientific concepts where they don't belong. So, your opponent may try to justify some spiritual mumbo-jumbo by appealing to quantum physics. See, for example, this quote from Deepak Chopra:

> *Quantum healing is healing the "bodymind" from a quantum level. That means, from a level which is not manifest at a sensory level. Our bodies ultimately are fields of information, intelligence and energy. Quantum healing involves a shift in the fields of energy information, so as to bring about a correction in an idea that has gone wrong. So quantum healing involves healing one mode of consciousness, mind, to bring about changes in another mode of consciousness, body.*

In such a case, you should ask, politely but firmly, how these concepts actually support your opponent's crazy inferences. Even better, you might ask how many quantum physicists also believe in quantum healing.

This fallacy embraces nearly every other fallacy there is. It derives from the Latin 'does not follow': the conclusion simply *does not follow* from the premises.

Interestingly, not every *non sequitur* is logically invalid. Logical validity is defined simply in terms of the truth of the premises and the conclusion: the premises cannot be true while the conclusion is false. So, an argument with false premises may still be a logically valid argument; an argument with logically necessary premises and conclusion will still be logically valid, even if the premises and conclusion have nothing to do with each other.

PROVING NON-EXISTENCE

Informal	Proponent A argues that P exists, because there is no evidence that it *doesn't* exist.

Asserting that something exists, on the grounds that its non-existence cannot be proven.

EXAMPLE

"I believe that ghosts exist. After all, I've never seen any conclusive evidence that they don't *exist!"*

REAL-LIFE EXAMPLE

Bertrand Russell once confronted the argument that belief in God was rational, since there was no evidence that He didn't exist. Lord Russell responded that, on this line of reasoning, one should also believe that there is a small teapot orbiting Jupiter: since there's no evidence that there *isn't* such a teapot, we should believe that it exists.

THE MISTAKE

This argument falsely shifts the *burden of proof* to the person denying the existence of an entity; however, the burden of proof should always fall on the person making the *assertion* not the refutation. To have a *guarantee* that something exists, I need *positive evidence* that it exists. This evidence can be quite low-grade (I believe there's an elephant in the room, because I see a huge grey pachyderm trampling over my furniture), or extremely sophisticated (I believe gravitational waves exist, because extremely complex scientific

experiments confirm their existence). But *some* sort of evidence, empirical or not, there must be: it's the job of the believer to convince the skeptic, not the other way round.

THE COMEBACK

Russell's teapot example provides an extremely effective *Reductio ad Absurdum* (see p. 165). If your opponent offers no evidence supporting the belief that his favored entity exists, and instead challenges you to show that it *doesn't* exist, then you can challenge him back: "I believe there is a teapot orbiting Jupiter. I see no evidence that it doesn't exist. So you should believe it too!" If your opponent has any sense, he should see the point immediately. If he ends up believing in the existence of such a teapot, I can't help you any further.

SIGNIFICANCE

This move is generally fallacious, because it falsely shifts the *burden of proof*. Your opponent believes, falsely, that it is the job of the skeptic to give reasons for why something *doesn't* exist, rather than incumbent upon the believer to give reasons why it *does* exist. But this move is wrong: it's generally the job of the believer to give reasons to believe that something exists. Of course, if he can do so, he has legitimately shifted the burden of proof: we should believe that P exists as he asserts, unless we can prove otherwise.

This fallacy is linked to the *Argument From Ignorance* (see p. 83), when it is asserted that something exists, or is true, just because there is no evidence that it doesn't exist.

RED HERRING

Ignoratio Elenchi, the Chewbacca Defense

Informal	Proponent A and opponent B are arguing about a topic P. B raises topic Q, on the grounds that it is relevant to P; however, Q is actually irrelevant to P.

Attempting to derail an argument by bringing in considerations that are irrelevant or out-of-context.

EXAMPLE

"My opponent criticizes China's oppression of Tibet. But what about the US occupation of Iraq? Isn't he concerned with that?"

REAL-LIFE EXAMPLE

A *South Park* episode ("Chef Aid") sees the eponymous Chef being sued by a major record company. The defense bases its case on (inaccurate) *Star Wars* trivia, stating:

> *Ladies and gentlemen of this supposed jury, I have one final thing I want you to consider. Ladies and gentlemen, this is Chewbacca. Chewbacca is a Wookie from the planet Kashyyk. But Chewbacca lives on the Planet Endor. Now think about it; that does not make sense! . . . If Chewbacca lives on Endor, you must acquit! The defense rests.*

What do these ruminations about Chewbacca have to do with Chef's harassment lawsuit? Nothing: that's the whole point. The

lawyer is deploying a *Red Herring* to disguise the fact that he doesn't have an argument (he still wins).

THE MISTAKE

Your opponent has brought in evidence that is simply *irrelevant* to the case at hand. It's as simple as that: what he's said is just beside the point.

THE COMEBACK

In an ideal world, you should be able to simply point out the irrelevance and move on. However, this tactic can be used in an underhanded way to completely derail the debate. So, your opponent might throw in a whole bunch of irrelevant considerations, leaving you with a dilemma: respond to his arguments, and you've wasted your time; fail to respond, and your opponent looks like he's scored a hit.

SIGNIFICANCE

A related fallacy is known as the *Ignoratio Elenchi* (Latin for 'ignorance of how something is refuted'): a situation where a (perhaps valid) argument is presented that is, however, beside the point. Aristotle characterized it succinctly: to refute an assertion, you must show that its *contradictory* is true; an *ignoratio elenchi* occurs when anything else but that has been demonstrated.

It is popularly supposed that the fallacy gets its name from the practice of using *red herrings* (kippers) to throw hounds off the scent of their quarry. However, this alleged etymology is misleading (a red herring, if you will): although red herrings *were* used in hunting, their purpose was not to distract the hounds, but rather to train the *horses* to *follow* the scent.

REDUCTIO AD ABSURDUM

Informal	Proponent A puts forward proposition *P*. Proponent B attacks a simplified or absurd version of proposition *P*.

Attempting to refute your opponent's argument by drawing allegedly absurd consequences from his argument, which, however, only follow from a caricatured misrepresentation of his position.

EXAMPLE

"You want the drinking age to be lowered from 21? Why don't we also let toddlers tend bar?"

REAL-LIFE EXAMPLE

Arthur Conan-Doyle (of Sherlock Holmes fame) wrote a lesser-known story called *Behind the Times*, concerning one reactionary Doctor Winter, who regarded Darwinism as

"the crowning joke of the century. 'The children in the nursery and the ancestors in the stable,' he would cry, and laugh the tears out of his eyes."

In other words, if Darwinism were true, then why not say that our ancestors were horses? (Other anti-evolutionist theorists resort to the same technique, with no more justification).

THE MISTAKE

This argument can come in a variety of forms, all of which involve attacking a *straw man* (an inaccurate caricature of a position). First, your opponent may try to present your position as far *stronger* than

it is, and attack this stronger argument as absurd–such is the case with the drinking-age example. Secondly, your opponent may try to draw absurd consequences from this straw man he concocted–as in the Dr. Winter example, who thinks that Darwinism implies that our ancestors were horses.

THE COMEBACK

You need to clarify your position, making clear that your argument isn't nearly as strong as your opponent suggests, or that he is missing some crucial subtleties in your position. So, with the drinking example, you can simply respond: "No, I'm not arguing that kids should tend bar; all I'm arguing for is that kids old enough to serve in the military, pay taxes or go to college, should be allowed to drink in a bar." With the evolution example, you could say: "No, I'm not saying that horses were our ancestors; but go back far enough and you will find a common ancestor that both we and they descended from."

SIGNIFICANCE

The *reductio ad absurdum* (Latin for 'reduction to the absurd') is a completely legitimate device in mathematics and philosophy, where it proves one thing by showing that its negation leads to absurdities. Thus, we know there is no rational square root of two: that is, no two integers *a* and *b* such that *a/b* is the square root of two. We know this because, if there *were*, either *a* or *b* would have to be *both* odd *and* even (this discovery, as we saw while discussing the *Moralistic Fallacy* [see p. 139], was the undoing of poor Hippasus). The reduction to the absurd only gives rise to a fallacy when the absurdity it exposes is predicated on misrepresentations of your opponent's position.

This argument can involve the *Appeal To Ridicule* (see p. 78) (as with the bartending example); it also encompasses the *Slippery Slope* (see p. 175).

REDUCTIO AD HITLERUM

Informal	Proponent A puts forward position *P*. Opponent B retorts that Hitler believed in *P*; therefore, we should not believe in *P*.

Dismissing your opponent's position on the grounds that Hitler (or some other evil figure) believed in it; or that the policy he advocates was also advocated by the Third Reich.

EXAMPLE

> *"You're a* vegetarian? *Don't you know* Hitler *was a vegetarian?!"*

REAL-LIFE EXAMPLE

There are so many examples of this that one is rather spoiled for choice. Liberalism, socialism, gun control, environmentalism, the traditional family, anti-smoking laws, exercise regimes: all have been linked to the Third Reich. However, we could make a special mention of Theodor W. Adorno's classic analysis of horoscope columns, *The Stars Down to Earth*. Adorno castigated a horoscope's advice to "accept all invitations" with the rebuttal:

> *The consummation of this trend is the obligatory participation in official 'leisure-time activities' in totalitarian countries.*

THE MISTAKE

This is an instance of an *Ad Hominem: Guilt by Association* (see p. 22), and is as risible as any other example of that fallacy. An

argument isn't necessarily wrong just because evil people happen to believe in it; this rule holds even if the evil person in question is Hitler himself. After all, we don't doubt that Berlin is in Germany just because Hitler also believed this; and we shouldn't refrain from building highways just because Hitler also did so.

THE COMEBACK

This fallacy admits of a quick and easy comeback: simply point out that your opponent's argument is irrelevant. Facts are not rendered false just because someone evil also believes in them; an argument isn't rendered invalid just because Hitler believed in it. You could then point out that Hitler believed that the Earth goes round the sun: does your opponent want to reject that proposition, on the grounds that Hitler believed in it?

SIGNIFICANCE

The name of the fallacy was coined by the political philosopher Leo Strauss, in an article published in 1951 in *Measure: A Critical Journal.*

This fallacy is so common in online debates that it has earned the sobriquet 'Godwin's Law,' after the academic Mike Godwin, who asserted that

> *"As an online discussion grows longer, the probability of a comparison involving Nazis or Hitler approaches 1."*

The corollary here is that the first person to mention Hitler or the Nazis has automatically lost the debate. This corollary, of course, must have certain limits: if one is discussing German domestic and foreign policy between 1933 and 1945, it would be inappropriate to invoke Godwin's Law.

SELF-SEALING ARGUMENT

Informal	Proponent A asserts a *substantive claim P*, such that no evidence can count against *P*, or that no opponent may raise an objection to it.

A substantive claim *which its proponent presents in a way that admits of no refutation, either by preventing any evidence from counting against it, or by automatically dismissing the objections of an opponent.*

EXAMPLE

"You're an alcoholic. What? You deny *that? That just proves it: resorting to denial is absolutely typical of alcoholism."*

REAL-LIFE EXAMPLE

Religions, Marxist schools, and schools of psychoanalysis have all been guilty of this fallacy. If you object to a certain religious doctrine, your position is summarily dismissed, since you are a *heretic*. If you object to a psychoanalytic doctrine, it's because you're employing psychological defenses against its uncomfortable truth. If you dismiss Marxism, it's because you are a lackey of the bourgeoisie. And so on.

THE MISTAKE

If you're making a substantive claim—a claim that's not just true by definition, that is, an irrefutable *logical* truth—then you are making a claim that *could* be false. As such, you need to think about the circumstances in which you'd admit that your argument is wrong: what evidence would you need in order to accept that you were in error. But a *self-sealing* argument rejects this stricture: its proponent simply won't accept that any evidence *could, even in principle,* make it incorrect. But such a stance is simply willful blindness: if you were wrong you'd have no way of knowing this, because you've insulated yourself from any criticism.

THE COMEBACK

Since this argument is the preserve of the fanatic, a comeback is unlikely to have any effect: after all, your opponent will only dismiss you as a heretic, a neurotic, or a reactionary running dog of the capitalist order. You can convince your *audience*, however, that your opponent's argument simply falls outside the realm of rational discourse, and therefore is not worth debating.

SIGNIFICANCE

A *self-sealing* argument is *unfalsifiable*: no argument may be raised against it. It is, at times, an *ad hominem* argument; in this case, however, the *ad hominem* extends to *any* individual who would doubt the argument. At other times, it can seem like *Cherry-Picking* (see p. 99): as far as your opponent is concerned, only such evidence as supports his conclusion counts. Any way you slice it, it's a lousy argument.

SHOEHORNING

Informal	A and B are discussing topic P. B (as is his wont) raises topic Q, where Q is irrelevant to P.

Where a contributor to an argument derails *the discussion by raising a favored topic of his, despite this topic's being completely irrelevant to the argument at hand.*

EXAMPLE

"Let me tell you about my views on Jesus. Jesus is my personal savior . . ."
"But we were discussing farming!"

REAL-LIFE EXAMPLE

Cato the Elder, a Roman senator, was an implacable foe of the city of Carthage, wishing that Rome's enemy would be wiped from the face of the earth. He was known for ending his speeches with the catchphrase *Carthago delenda est* ("Carthage must be destroyed"), even when discussing quite unrelated matters; as in, "the corn grows high, and Carthage must be destroyed."

THE MISTAKE

This fallacy is typically one of *irrelevance*: you are discussing one topic; your opponent has brought in considerations that have nothing to do with the topic at hand. Although this isn't technically a *fallacy*, because it doesn't really engage the argument at hand, it is, however, a tedious maneuver.

Just ask your opponent what the relevance of his contribution is to the topic at hand. He may dearly love to discuss Jesus with you, but since your topic is farming, his contribution is quite beside the point.

SIGNIFICANCE

This isn't a *fallacy* per se: it's not that your opponent is trying to draw a conclusion from premises that won't allow it; rather, your opponent is introducing an out of context topic . Your opponent's argument may in fact be cogent, valid, well-articulated; however, his contribution is pointless if it doesn't add anything relevant to the substance of the discussion.

Shoehorning tends to be habitual, rather than a one-off behavior: the person who *shoehorns* tends to have a hobby-horse which he tries to bring into the discussion as often as possible, even out of context. Such people are best avoided, as they tend to be bores.

SLIPPERY SLOPE

Absurd Extrapolation, Camel's Nose, Thin End of the Wedge

Informal	If *A*, then *B*; if *B*, then *C*; if *C*, then . . . *Z*!

Predicting that horrific consequences will follow from seemingly innocuous actions, through an incremental, step-by-step process. So, if we do A, *this will inevitably result in action* B, *which will result in* C . . . *which will result in (typically horrific) action* Z.

EXAMPLE

> "If the government bans students taking guns to school, next it will ban any guns being taken to school. Next, it will ban anybody taking guns to any public places. Before you know it, the government will ban guns altogether!"

REAL-LIFE EXAMPLE

> "Finally told me, said: I don't like the way this country is headed. I want my granddaughter to be able to have an abortion. And I said well mam . . . [t]he way I see it goin' I don't have much doubt but what she'll be able to have an abortion. I'm goin' to say that not only will she be able to have an abortion, she'll be able to have you put to sleep" (Cormac McCarthy, No Country for Old Men).

In other words, legalize abortion, and involuntary euthanasia will inevitably follow.

This fallacy, as the name suggests, tries to show that horrendous consequences will *necessarily* follow from innocuous decisions, via a series of small, incremental steps. The fact is that they don't have to: there is simply no reason to think that accepting the first course of action will lead to another; still less, that it will lead to the awful ultimate consequences envisioned. In reality, we can usually stop the chain of consequences at any juncture.

THE COMEBACK

Challenge your opponent to show *how* taking the first step in this chain of consequences will lead to the horrific consequences he predicts: why, when we take the first step, can we not simply stop there? Or, challenge him by showing that other considerations will prevent us from taking these further steps. For example, if he argues that legalizing gay marriage will lead to legalizing incestuous or inter-species marriage, you might respond that it doesn't have to: even if gay marriage is legalized, these further expansions of marriage would be ruled out, because of the prohibitions on incest and bestiality.

SIGNIFICANCE

The name *slippery slope* illustrates the logic of this fallacy well: once you're at the top of a slippery slope, it's very hard not to slide down, all the way to the bottom; the best thing, then, is not to get on the slope at all.

Slippery slope arguments *do* have their place, provided that we can spell out exactly how the first step will lead to all the others. So, the first step might change people's attitudes sufficiently that they will more easily accept the next steps; or, it may be that people would object to change brought in all at once, but not if it's brought in incrementally. This is best illustrated in Niemöller's poem: "*First they came for the Jews . . .*"

SPECIAL PLEADING

Informal	Proponent A agrees to a general rule P. P applies to B. A demands that an exception be made regarding P's application to B, without giving grounds as to why an exception is warranted.

Agreeing to a general rule or principle about something, only to suspend it in a particular instance, without giving any good reasons for doing so.

EXAMPLE

> *"All able-bodied men in this state should go to war! All, except my son . . ."*

REAL-LIFE EXAMPLE

In *Downton Abbey*, the butler Molesley is called to war. Cousin Violet, the indomitable Countess of Grantham, colludes with Dr. Clarkson to try and get an exemption for him; however, Mrs. Crawley, who runs the local hospital, uncovers the deception. Cousin Violet pleads: "But he is my gardener's only son!" to which Mrs. Crawley coolly replies to this effect: "But *everybody* is someone's son. I am sorry, but if I make an exception in this case, I would be forced to make an exception in every case."

THE MISTAKE

If you agree with a general law, then you have to agree with it in all cases. If you find that it cannot apply to certain circumstances,

then you disagree with the law; else, you need to refine the law, or perhaps demonstrate that the spirit of the law does not warrant its extension to this case. What you cannot do, however, is accept the law in general and yet refuse to apply it to a particular case. To do so is essentially self-contradictory. Worse, once you go down that route, there's no reason why *other* people shouldn't also be allowed to make ad hoc exceptions: if *your* son should be allowed to refuse military service, why shouldn't *mine*?

THE COMEBACK

You need to impress on your opponent the virtue of consistency. If your opponent is committed to a rule, he must be committed to it without exceptions. If he believes there are exceptions, he's not committed to the rule. Further, you should remind him that, if we start making exceptions, then there's no reason not to turn *every* case into an exception. This is an instance where invoking the *Slippery Slope* (see p. 175) becomes a legitimate countermove.

SIGNIFICANCE

This fallacy is sometimes given a fig leaf of support by applying ad hoc justifications to the rule ("But you let your dog in!" "Well, it says no dog*S* allowed: we're allowed to have *one*!"). Such a move, however, is usually a pretty transparent dodge: you can always ask *why* such a rule-patch is needed. An adequate justification will usually not be forthcoming.

SPIRITUAL FALLACY

Informal	Proponent A makes claim *P*. *P* does not come about. Proponent A claims that, despite appearances, *P* has actually come about "in a spiritual sense."

Taking a claim (usually a prediction) to be satisfied, despite lack of visible evidence for it, by asserting that it has been satisfied in a "spiritual sense."

EXAMPLE

"Our beloved cult leader has returned, to shower his eternal benevolence upon us!"

"What are you talking about? He's dead! He's tagged in the morgue!"

"Ah, yes, but don't you see? His mortal vessel may be dead, but he has returned to us in a spiritual sense!"

REAL-LIFE EXAMPLE

In Evelyn Waugh's classic novel *Brideshead Revisited*, Rex Mottram is undergoing the Catechism to convert to the Catholic faith. His instructor recounts their class thus:

> Then again I asked him: "Supposing the Pope looked up and saw a cloud and said 'It's going to rain,' would that be bound to happen?" "Oh, yes, Father." "But supposing it didn't?" He thought a moment and said, "I suppose it would be sort of raining spiritually, only we were too sinful to see it."

THE MISTAKE

Your opponent has been caught with his pants down: he's made a claim that hasn't come to pass. His response, namely that his claim has come true 'in a spiritual sense,' is clearly a desperate move. But not only is it desperate, it is also epistemically problematic. How on earth would you know that such a claim is true, even in this very nebulous sense? And, equally importantly, how would you know if it were *false*? (see *Self-Sealing Argument* [see p. 171] and *Unfalsifiability* [see p. 189]).

THE COMEBACK

This move doesn't really deserve a comeback. You could give your opponent a stern lecture about why substantive claims must have some falsifiability condition, or why self-sealing arguments are bad; however, it would probably have little effect on him.

SIGNIFICANCE

This fallacy is related to *unfalsifiability*: your opponent argues in such a way that nothing could prove him wrong. If Papal Infallibility did mean (it does not) that everything the Pope said was bound to be true, we could test it easily: if the Pope says it will rain, and it does not, then the doctrine would have been *falsified*. But if it's true, because it's 'sort of raining spiritually,' then the doctrine is *unfalsifiable*: how would you know if it was false?

This fallacy is also a form of *Moving the Goalposts* (see p. 142). So, the claim goes, my prediction hasn't come true in a literal sense; however, it *has* come true in a much more nebulous sense.

STRAW MAN ARGUMENT

Informal	Proponent A puts forward argument *P*. Opponent B rebuts *P* by actually rebutting *P'*, which is superficially similar to, but importantly different from *P*. Opponent B takes his rebuttal of *P'* as a refutation of *P*.

Misrepresenting an opponent's argument, directing your attack at the misrepresentation, and taking this attack to refute your opponent's real position.

EXAMPLE

> *Someone says: "I think we should do more to conserve the environment, and not cover pristine countryside with freeways."*
>
> *And another person responds: "So you don't want freeways to be built at all? How on earth do you imagine we'd get around?"*

REAL-LIFE EXAMPLE

Some arguments against evolution rely on this trope. So, the classic rebuttal 'if evolution is true, how come there are still monkeys?' relies on a misunderstanding of evolutionary theory. Evolutionary theory neither teaches that humans evolved from monkeys (they share a common ancestor), nor that the evolution of a new species necessarily entails that the old species go extinct (the new species could evolve, for instance, if a sub-group of a species migrates to a new environment, and evolves in response to novel environmental pressures). The rebuttal is effective only when attacking a grossly simplified and inaccurate version of evolutionary theory.

THE MISTAKE

It should go without saying that, to effectively rebut an argument, you have to actually rebut *that very argument*. It's no good arguing against something that's sort of similar to it: unless your rebuttal is directed against the argument itself, you won't get very far. The *Straw Man Argument* is a particularly pernicious example of this problem, since it argues against a position by arguing against a simplified misrepresentation of it, ignoring the various qualifications and subtleties essential to the position.

THE REBUTTAL

You have to call your opponent out for misrepresenting your position, and remind him what your position actually is. You should then take pains to point out how, when confronted with your *actual argument*, his supposed rebuttal cuts no ice. So, in the above examples, Person A has not argued against building freeways in general, but only against despoiling pristine countryside; however, Person B's refutation ("how will we get around?") really only works if Person A is arguing for the stronger claim. Similarly, the rebuttal 'how come there are still monkeys?' works against the Darwinist only if the Darwinist holds a grossly simplified version of evolution. Point out what you *actually* believe, and show that your opponent's argument is ineffective.

SIGNIFICANCE

This fallacy is related to the *Reductio ad Absurdum* (see p. 165) (where the opponent's argument has apparently absurd consequences drawn from it), as well as the *Slippery Slope* (see p. 175) fallacies (where the opponent's argument is shown to have pernicious consequences).

SUNK COST

Informal	Investor A has sunk *n* units of currency into project P. Although P has little chance of making money, A continues to sink money into it, because he does not want to give up on *n*.

Sunk costs *are the resources invested in a project or venture which have become irrecoverable by any means. The* Sunk Cost *fallacy occurs when the investor continues investing money in a project, despite having little or no hope that it will make a return above funds already invested, because of a reluctance to let the initial investment go.*

EXAMPLE

John buys five raffle tickets at $1 each, hoping to win a prize worth $10. His numbers having failed to come up, he buys a further ten tickets, to be in with a chance of winning the prize; however, he could simply spend the $10 to buy the prize item.

REAL-LIFE EXAMPLE

Stephen Colbert used this type of reasoning ironically, pretending to justify the United States' continued use of torture during the War on Terror. He reasoned that, because the US had already lost its position of moral superiority, it should continue torturing terrorist suspects, until it could be shown unambiguously that torture saves lives. Or, as he put it:

The point is, we know there is no going back. We have already tortured people. You cannot un-ring that bell. (Or Un-Wring Those Balls). And the knowledge that we have already tortured people could be the death of America's moral leadership. So to save that, we must do whatever it takes to justify what we've already done.

THE MISTAKE

Sinking more money into an unprofitable venture is simply *irrational*. The resources the investor put into the initial venture are, as the name suggests, *sunk*: there is no way of recovering them. As such, the only relevant consideration as to whether to invest more resources is whether *these additional* resources would have a realistic prospect of return. In the *Sunk Cost* fallacy, however, the previous investments are erroneously included in this calculation.

THE COMEBACK

You need to remind your opponent that the capital he's invested in the venture simply cannot be recouped. Past losses are therefore irrelevant to the decision at hand: the only relevant consideration is whether it would be worth investing any *additional* capital in the venture. Undoubtedly, it is regrettable that you've already sunk your resources into such a project; it can be hard to accept that these resources are simply gone. But doing so is nevertheless better than the alternative, i.e. to simply keep pouring time or capital into a venture which has no chance of success, when the likely outcome will simply be that *these additional* resources will *also* go to waste.

SIGNIFICANCE

This fallacy isn't so much a problem of logic, as of psychology and economics: we are prone to miscalculating the significance of resources that we've already invested into a venture. This sometimes comes about because we are scared of losing face: to cut your losses and run would be to admit that you've made a mistake. But such misguided pride has grave consequences, whether it concerns the gambler who bets his house on the roulette wheel, or the country that keeps sacrificing its troops in order to not give up fighting an unwinnable war.

UNFALSIFIABILITY

Formal	Proponent A makes a claim that *P*, such that there is no way of disproving that *P*.

A substantive proposition is expressed in such a way that it becomes, in principle, impossible to raise a counterexample to it.

EXAMPLE

"God answers all my prayers!"

"But you prayed that Aunt Edna would get better from cancer; yet she died."

"Yes, so God clearly answered my prayers, and let Aunt Edna pass to the next world, which is a better place."

REAL-LIFE EXAMPLE

According to classic Marxist theory, the development of capitalism would see the proletariat become increasingly impoverished, which would lead it to organize as a class and eventually overthrow the wicked bourgeoisie, thus ushering in a socialist paradise. This theory's predictions, however, were somewhat nebulous; the theory did not offer a criterion by which they could be judged to be *false*. Marx proposed no crucial experiment, for example, by which the truth or falsity of his predictions could be ascertained. As a result, despite Marx's predictions not having come to pass, his theories still retain a number of adherents.

If a claim is a substantive one, it could be true, but it also could be false. As such, you need some sort of criterion by which you can judge that it is either: if *this* happens, the theory is true (*verifiability*); if *that* happens, it is false (*falsifiability*). If such an unequivocal criterion is not forthcoming, the theory is *unfalsifiable*; however, if it is unfalsifiable, then there is simply no way of telling whether the theory is false or not.

This mistake can take several forms. At its most basic, your opponent may not be able to propose any criterion by which his theory might be judged false. However, it may also be the case that your opponent takes any evidence whatsoever as corroborating his claim (as with the prayer example above); or that he explains his theory's failure by some ad hoc maneuver (e.g. "The proletarian revolution did not come to pass because mass culture has inured the workers from the realization of their own misery").

Ask your opponent as to what, exactly, would take him to admit that his theory were false: can he propose, for example, an experiment by which his theory may be put to the test? If he can't, then his theory is deficient: how would you *ever* know if the theory is false?

Falsifiability was emphasized by Sir Karl Popper as a *demarcation criterion* for science. For Popper, science was distinguished from pseudoscience on the grounds that the former admits of falsification, whereas the latter doesn't. So, the scientist, proposing that all swans are white, makes a *falsifiable* statement: show him a black swan, and he will accept that his theory needs rethinking. By contrast, the pseudoscientist's assertions aren't like that: he can't tell you, even in principle, what sort of thing would make him retract his statement.

This demarcation looks pretty clear in the abstract; nevertheless, it can be much harder to apply in practice. Scientific theories very seldom admit of a quick and easy refutation: if the facts don't fit the theory, the theory can be tinkered with, instead of outright discarded.

USE-MENTION ERROR

Informal	Proponent A discusses word 'P.' Opponent B thinks that A is discussing the concept or object that P denotes. Consequently, confusion arises.

Confusing the discussion of a word itself with discussing the concept the word denotes.

EXAMPLE

"My son is Adam. 'Adam' has four letters. Therefore, my son has four letters."

REAL-LIFE EXAMPLE

In *Monty Python's Life of Brian*, an unfortunate Judean called Matthias is condemned to death by stoning, for uttering the word 'Jehovah' during dinner. The execution scene plays out thus:

MATTHIAS: Look. I'd had a lovely supper, and all I said to my wife was "That piece of halibut was good enough for Jehovah."
CROWD: Ooooh!
OFFICIAL: Blasphemy! He's said it again!
CROWD: Yes! Yes, he did! He did!
. . .
MATTHIAS: Look. I don't think it ought to be blasphemy, just saying "Jehovah."
CROWD: Oooh! He said it again! Oooh!
OFFICIAL: You're only making it worse for yourself!
. . .

OFFICIAL: *Stop! Stop, will you?! Stop that! Stop it! Now, look!*
No one is to stone anyone until I blow this whistle! Do you
understand?! Even, and I want to make this absolutely clear,
even if they do say "Jehovah."

At this juncture, the Official is promptly stoned.

Matthias is accused of having uttered Jehovah's name over dinner, which is deemed a blasphemous act. However, in the following scenes, he merely *mentions* the word Jehovah: he is referring not to Jehovah, but to the *word itself*; the word's *use* is discussed, but is not itself *used*. So, in this scene, neither the Judean nor the Official actually *use* the word "Jehovah" (in the sense of denoting God), they only *refer* to it. Not that it matters, of course: the Official still gets stoned by the angry mob.

THE MISTAKE

Words are useful, because they denote things that aren't them– this is when a word is *used*. However, we sometimes want to refer to a word itself: this is when the word is *mentioned*. The difficulty arises when these different activities are conflated: when talking about the name itself is taken as equivalent to talking about the thing it denotes, or when talking *about* the name is taken as an instance of *using* the name. I can say things about a word that are not true about the thing it denotes (the statement "'Adam' has four letters" does not imply that *Adam* the person has four letters), and things about the thing (the word's referent) which are not true of the word (that Adam is blond does not mean that the word "Adam" is blond).

THE COMEBACK

Your best option, in this case, is to simply remind your opponent of the distinction, namely that talking *about* a word is not tantamount to *using* the word as such. Alternatively, if he is particularly recalcitrant, you could trick him into saying the word itself, as in the case of the Official in the above example.

SIGNIFICANCE

This fallacy is sometimes encountered (as with the Jehovah example above) in the context of taboo words. So, if I say "damn," then I've sworn, because I've *used* a swear word; if I say "John said 'damn,'" I haven't sworn: I've just *mentioned* the swear word. But a word often retains its taboo function, even when it's simply mentioned. That is why, for instance, we speak of the "F-word" rather than spelling it out.

Index